Pharmaceutical Discovery, Development and Manufacturing Forum 2021

Held at the 2021 AIChE Annual Meeting

Boston, Massachusetts, USA and Online
7 - 11 November and 15 - 19 November 2021

ISBN: 978-1-7138-5719-8

Printed from e-media with permission by:

Curran Associates, Inc.
57 Morehouse Lane
Red Hook, NY 12571

Some format issues inherent in the e-media version may also appear in this print version.

AIChE
120 Wall Street, FL 23
New York, NY 10005-4020

Phone: (800) 242-4363
Fax: (203) 775-5177

www.aiche.org

Additional copies of this publication are available from:

Curran Associates, Inc.
57 Morehouse Lane
Red Hook, NY 12571 USA
Phone: 845-758-0400
Fax: 845-758-2633
Email: curran@proceedings.com
Web: www.proceedings.com

TABLE OF CONTENTS

ACCELERATING TIME TO AUTONOMY IN AN INDUSTRY 5.0 WORLD

Brent Kedzierski, HumanWRKS LLC, Houston, TX
Ian Willetts, AVEVA Software LLC, Carlsbad, CA

Introduction

Industry 4.0 is about smart production, IoT, sensors, drones and other cyber physical innovations. Industry 5.0 will be about connecting smart production to human cognition, collaboration, and creativity. To this end, progress will be judged on not just the connection of data between the Internet of Things (IoT), but how that connectivity enhances the human experience and shortens the time to achieve real autonomy for today's connected worker.

This paper discusses how AVEVA are working together with international oil companies (IOCs) to develop and deploy more personalized worker experiences to serve the new industrial agenda of achieving the goals of both, the enterprise as well as continually ensuring the welfare of humans.

Industry 4.0 Needs People with Connected Worker Skills

Industry 4.0 emerged in 2011 and has been the most technologically sophisticated and focused of all industrial revolutions. Its focus has been on transforming industrial manufacturing through digitalization and the exploitation of the potentials of new technologies. The suite of connected technologies that have been adopted across industry over the last decade are commonly referred to as the 9 pillars of Industry 4.0[1] (Simulation, Augmented Reality, Big Data, Additive Manufacturing, Cloud Computing, Autonomous Robots, Cybersecurity, IIOT, Horizontal and Vertical System Integration). These pillars seek to work together as disruptive connectable technologies to transform industrial production by creating an industrial ecosystem that is both automated and intelligent.

However, research reveals that many manufacturing leaders believe there are a lack of skilled workers across industry to support their digital transformation goals[2]. The skills gap is understandable given that confidence and adoption levels of automation and digital transformation innovations are typically spread across a bell curve of innovators, early adopters, early majority, late majority, and laggards. The second decade of Industry 4.0 will see mature connected ecosystems become common place across major industries. However, the lack of requisite connected workers will have a dual effect on the advancement of smart industries:

- First, industry will experience a value degradation in their implementations of advanced technologies without enough autonomous and digitally fluent workers.
- Second, the industrial workforce will become increasingly disengaged as opportunities to grow their skills are superseded by a focus on automation, resulting in issues related to workforce attraction, retention and development.

Workforce competence development in the past has always been understood to be a critical enabler of operational excellence and safe operations. However, as connected technologies mature industry must take a long hard look at developing a highly competent and autonomous workforce in parallel to the deployment of their Industry 4.0 initiatives.

Shift from Competence to Autonomy

Across industry the phrases "time to competence" and time to autonomy" have often been used interchangeably. However, autonomy is a very individual state of being that goes beyond an organization simply giving an employee the license to act based on the assurance of their competence.

To provide context on the difference between competence and autonomy we can consider a driver that is learning to drive a car. They will study the theory and take driving lessons on the road with a view to passing the driving test. The driver will reach a level of competence so they can pass the test and know "how" to drive a car. However, that driver will only become truly autonomous after an extended period of driving where they are comfortable in the seat and able to be aware of their surroundings, understand the cause and effects of their decisions and truly contribute to a safe driving experience for everyone.

Even when world class organizations such as Shell took a concerted effort to implement a global development program in 2011 the program was based on a traditional programmatic competence development and assurance program and not designed to enable true worker autonomy. The program was effective in reducing time to competence requirements, but competence alone does not equal autonomy. At best competence provides management with a single factor level of assurance that workers possess the requisite knowledge and skills linked to a standardized framework.

To succeed in today's tech heavy industrial settings and meet evolving worker expectations, industry must move past the traditional view of competence to drive towards a true level autonomy across its workforce. The advanced technologies of Industry 4.0 will require a more human centric framework for worker autonomy in order to truly enable the optimization of the human machine mix.

The actual tools of the Industry 4.0 revolution such as the increased availability of real time data, simulation, cloud computing and virtual reality can be leveraged to not only improve the manufacturing process, but they can also be used to accelerate worker engagement and time to autonomy for the connected worker.

Industrial Workforce Autonomy – Industry 5.0

There have always been six (6) essential components of workforce autonomy. Each one of these components fall under the three (3) core areas of human ability:

- Cognition
- Creativity
- Social Connectivity

The desired end state is ensuring the worker can consistently:

1. Execute job tasks, analyze, synthesize, and evaluate relevant job/task data
2. Apply sharp cognitive abilities to apply data to troubleshoot and make decisions

3. Apply their creative abilities to solve problems
4. Leverage their social skills to collaborate with colleagues in all areas of work
5. Know when and when not to escalate issues.

Looking at each of the above components of autonomy, industry must shift from a technology centric 4.0 viewpoint to a connected worker 5.0 viewpoint. Industry 5.0 will represent a human centric revolution where technology will play a secondary role in the hunt for industrial expansion.

The modern industrial worker must be equipped with the skills to thrive and work autonomously with intelligent technologies and data rich cultures to drive advances in innovation and sustainability. Industry 5.0 has entered the scene because industry, society and the types of gains required in today's economy are at a place where a value-based revolution is needed. Over the next decade industry will look within every nook and cranny of the industrial framework to reposition the human and the human-machine mix to deliver a new level of value for industrial stakeholders.

Strategies for Building Autonomy

The need for a more strategic development model for workers is now at a critical tipping point across industry. These strategies must work to develop humans that can not only co-exist but thrive in a tech heavy industrial world.

The jobs that are emerging due to connected technologies will have higher order intellectual, social, creative, and behavioral components that will present new challenges to human competence, efficacy and autonomy. Part of the new upskilling and reskilling dilemma is that traditional industrial skills must shift to match the demands of future connected industry. The opportunities across the entire industrial ecosystem to evolve new skills that derive autonomy can't be developed by osmosis or with subpar learning experiences as in previous decades.

In any effort to transform an organization, the first step is to establish a sense of urgency based on recognizing the need to change[3]. Also, complex, highly automated production systems leave fewer opportunities for on-the-job training, while also raising the stakes of human error in abnormal situations. These changing circumstances create an urgent need to improve training, but industry is slow to shift from traditional approaches. We also must recognize that today's workforce has very different expectations about technology which is important because technology can underpin new ways to drive toward autonomy.

From Training to Learning in the Flow

In their book, Minds at Work: Managing for success in the knowledge economy, David Grebow and Stephen J. Gill explain that organizations will need to shift from a training culture to a learning culture[4]. Training is the giving of information and knowledge, through speech, the written word or other methods of demonstration in a manner that instructs the trainee. Learning is the process of absorbing that information to increase skills and abilities and make use of it under a variety of contexts.

Learning is aligned with results. Learning-centered organizations make information accessible and individually relevant. Competence is judged by one's demonstrated abilities, not by an accumulation of seat time. Continuous learning in areas that matter to worker identity is key. People constantly create, share, and adopt information to make the company smarter in a learning organization.

Simulation Based Experiential Learning

Learning by doing has been a key mantra in the learning field for decades. Good simulation-based learning designs have been based on ensuring a few basic instructional practices are embedded into learning design. These included elements such as ensuring a clear goal, providing relevant context, challenge, giving learners choice options, showing consequence of actions, and coaching or feedback support. According to the NTL Institute for Applied Behavioral Science[5], simulation-based experiential learning produces more than seven times the knowledge retention of other training methods5.

In building autonomy, the goal is to build worker efficacy in their freedom to make choices for themselves rather than relying on others. Simulation based learning provides a unique learning experience where the decisions made by the learner on a simulator, during a scenario-based exercise, creates a realistic feedback loop where the consequences of an action, good or bad is seen and understood. The context rich nature of simulation also creates a unique learning experience for every single person, every lesson. There is no better way to build autonomy than by seeing and learning from the consequences of your actions in real time.

Immersive Training Simulators (ITS)

Immersive simulation tools use augmented reality and virtual reality to connect control room operators, maintenance, and field personnel in a single realistic learning environment. It is broadly used for site familiarization, work policies, field operations training and maintenance procedures. Augmented reality can be especially useful for remote collaboration with experts.

While all people are different and learn in different ways, simulation-based learning can help to cement lessons most effectively, using an immersive yet safe replication of real-world scenarios. Rather than simply knowing what they should do, immersive training allows users to see, hear, and feel what to do, so that when they are deployed in the real world, the right habits will have already been ingrained. One of the main principles of a learning-centered organization is to encourage measured risk-taking and experimentation. In an industrial environment, simulation is often the only safe, practical option for experimenting with risk-taking.

AVEVA has delivered to customers a full complement of behavioral learning modules utilizing the Cloud hosted AVEVA XR Virtual Reality platform.

Operator Training Simulators (OTS)

Operator error is a well-known problem. In the US alone, more than 84 incidents occur per week[6]. Chemical plants lose more than $20 billion annually from human error and insufficient knowledge[7]. A poorly trained operator can undo the most advanced engineering, placing a company's people and profit at risk.

An operator training simulator combines dynamic process simulation models with emulated control systems to create a virtual control room specific to a single plant or process. An operator training simulator is a building block of Shell's Multi-Purpose Dynamic Simulation (MPDS) initiative which uses the dynamic process model to support FEED and detailed design, the integrated dynamic model and controls database for virtual controls commissioning, establishing operating procedures and for operator training that provides management with confidence that that operators are ready for day one.

Eric Scappazzoni of the TOTAL Energies Manufacturing & Projects Division when interviewed said that he strongly believes using OTS is really improving TOTAL's operational excellence by increasing the competency and reducing the risk of having accidents. Other IOCs are currently executing programmatic deployments of Operator Training Simulators with AVEVA with a view to driving operational excellence and increased safety, while upgrading their learning experience for operators.

Historically OTS systems have been delivered on premise with a relatively large hardware footprint. Companies have realized the expected benefits of such systems in the early deployment period; however, a common challenge is that companies are often not willing to invest in long-term maintenance. AVEVA has addressed the challenge of the long-term maintenance of on-premise systems by providing the option to deploy both OTS and ITS systems to the Cloud.

Cloud Enabled – Anywhere, Anytime, Any Device

The optimum learning platform must leverage the Cloud pillar of Industry 4.0 to empower the connected worker to be autonomous in choosing when and where they invest their time and attention for their individual development. Today's simulators are more advanced than ever and bolstered by Cloud delivery that eliminates many of the traditional barriers to incorporating simulation-based learning into the curricula and serving the content up to the workforce anytime, anywhere and on any device.

This last point is paramount as the competition for learner's time and attention grows, the relevant factor will be where learners see the investment their time and attention paying the most dividend. Choice is about where the learner feels they can benefit most and in line with autonomy that means providing workers with greater control over both their work and their own learning and development.

Instructional Design

As the world of work and learning continue to blend closer together, it is important to consider models that factor in more human centric theories that are not only based on good learning design but are also relevant to driving autonomy. Stronger technology-based learning programs are much better equipped to incorporate higher concept and higher touch design elements that accelerate autonomy. These include considerations from human centric theories such as self-determination, self-efficacy, expectation value assurance theory, as well as attributes from human fulfillment and flow. Learning designs that factor in the elements of these theories will help shift industry from simply building competence against a static portfolio of skills to driving autonomy across a dynamic portfolio of worker enabled outcomes.

With the above platform that leverages three of the nine pillars underpinning Industry 4.0, organizational learning capabilities will mature to create content and context rich scenarios that will be less rote but instead more dynamic and challenging. Learning capabilities will mature to offer multi-model content that provides the learners:
More choices during each learning exercise

- The ability to analyze tasks and synthesize and evaluate relevant information
- The opportunity to evaluate conditions, make decisions and see "what if" consequences
- The chance to identify opportunities for hierarchical and matrixed collaboration

The learner is no longer watching video content and checking boxes. Instead, the worker is on a path to renewal and autonomy and all the benefits therein.

Unified Learning

Companies already invest substantial time and money into operator training. Simply investing more may yield some improvement, but it will not generate the transformation that the problem demands. In the Baker Report on the Texas City disaster, investigators noted that BP's computer-based training produced little retention or competence among operators, because the culture treated learning as just a box to check and then move on[8].

AVEVA's Unified Learning platform moves companies beyond point in time solutions such a rote e-Learning content such as that previously deployed by BP to a fully integrated Cloud hosted learning platform including:

- Instructional System Design methodologies
- E-Learning courses
- Process Competency Modules
- Generic Training simulators
- Custom High Fidelity Training Simulators
- Immersive Training Simulators
- AI Infused Microlearning
- Learning Management System

Rather than using disparate point solutions AVEVA's Unified Learning provides the optimum environment to allow companies to accelerate the time to autonomy for their workforce during this Industry 4.0 industrial revolution.

Conclusion

Without building more direct paths to higher levels of individual autonomy, industry will lack the ability, flexibility, and adaptability to optimize the connected technologies of Industry 4.0 in ways that fully harness and integrate the full spectrum of unique human capabilities.

Great companies will build human performance experience models that enable getting work done and harvest a core byproduct with opportunities for the worker to experience greater levels personal fulfillment and growth.

AVEVA is working diligently to ensure that its Unified Learning Platform provides all the necessary infrastructure and thought leadership to accelerate the time to autonomy for the workforce of today and tomorrow.

References

1. Wai Yie Leong, ; Joon Huang Chuah, ; Boon Tuan Tee, (ed.): 'The Nine Pillars of Technologies for Industry 4.0' (Telecommunications, 2020) DOI: IET Digital Library, https://digital-library.theiet.org/content/books/te/pbte088e

2. Hipplod, Sarah. Gartner Press Release. STAMFORD, Conn., May 11, 2021.

3. Kotter, J. P. (1995). "Leading Change: Why Tranformation Efforts Fail," Harvard Business Review.

4. Gill, S. J., & Grebow, D. (2017), "Minds at Work: Managing for success in the knowledge economy," Alexandria, VA: ATD Press

5. NTL Institute for Applied Behavioral Science

6. ARC Advisory Group

7. Abnormal Situation Management Consortium (2019)

8. Baker, J. A. et al (2007) "The Report of the BP U.S. refineries independent safety review panel"

622059 How to Get a Head Start on Your Digitalization Journey

Abstract:

Companies of all types and from all industries are increasing their productivity with digital technologies. With the right tools, they have the ability to increase efficiency, security, and productivity. They can achieve this by taking a holistic view of existing strategies and identifying smaller, more manageable process optimization opportunities of the existing installed base. In doing this, they need to leverage technologies such as artificial intelligence, smart monitoring and data analytics. The outcome is a customized digital transformation roadmap.

It is relatively easy to talk about the need for a digital transformation from a high level, but when it comes to planning and executing the strategy to create the specific roadmap, it tends to be more challenging. But it doesn't have to be that way. Industry 4.0 has now been around for a few years, which has allowed tried and tested approaches and methodologies to emerge that can be applied to pretty much any size company in any industry.

During this presentation you will gain the following insights:

- How to take the first steps of the digital transformation journey
- View of simulated business cases to dive deeper into the basic understanding of how to identify relevant, Industry 4.0 projects
- How to best prioritize these in terms of resources, return on Investment, digital maturity level, and more.

632003 Development of an Advanced Control Infrastructure for Subcritical Power Plant Online Power Demand Tracking

Abstract:

Many subcritical coal-fired power plants that usually operate in baseload conditions are migrating to load-following operations as more renewables are penetrating into the power grid. As renewables, such as solar and wind, deliver intermittent power, dispatchable energy systems such as coal-fired power plants are required to cycle their load to accommodate power demand fluctuations. However, most of the subcritical plants were not designed to operate under load-following conditions, resulting in additional emissions and increased power plant damage. Given that cycling may become the new operating mode for coal-fired power plants, there is a critical need to properly control these plants. In particular, to operate in different power dispatching modes, a power plant will need to produce steam at certain flows, temperatures and pressures as required by the turbine. The highly variable resulting plant dynamics will require predictive control algorithms capable of maintaining the process within the operating regions of interest.

In this work, an advanced control structure is developed for online control of the total power demand of a subcritical coal-fired power plant cycling around the setpoint of 200 MW. This structure involves the combination of proportional-integral-derivative (PID) and advanced model predictive control (MPC) algorithms [1] for real-time control of power demand. Specifically, temperatures of the primary and secondary desuperheaters are controlled by a 3 input x 3 output MPC that manipulates 2 spray valves and coal flowrate as inputs. Also, the PIDs (2) control the pressure and water level of the drum boiler by manipulating the water valve and steam-to-turbine valve, respectively.

A communication infrastructure encompassing the power plant model [2], real power plant data obtained from online databases (e.g., ERCOT [3]), and the control algorithms is established using the OSIsoft PI system at West Virginia University (WVU). The PI system centralizes the information received from the power plant model in MATLAB Simulink and the online power demand from the web, and send the control actions calculated by the MPC in MATLAB back to the power plant for implementation. To evaluate the proposed infrastructure, several days of power plant online data with cycling profiles are considered. The developed infrastructure enables power plant control using real-time data to improve decisions and operations of existing and future fossil-fuel-based power plants.

[1] He, X.; Lima, F. V. A modified SQP-based model predictive control algorithm: application to supercritical coal-fired power plant cycling. 2020. *Industrial & Engineering Chemistry Research*, *59*(35), 15671-15681.

[2] Agbleze, S.; Lima, F. V.; Indrawan, N.; Panday, R.; Pezzini, P.; Bonilla-Alvarado, H.; Bryden, K. M.; Tucker, D.; Shadle, L. J. Modeling and control of subcritical coal-fired power plant components for fault detection. *In Proceedings of the 2020 ASME ICONE 28-POWER*, August 2020.

[3] Electric reliability Council of Texas (ERCOT). (n.d.). Retrieved April 10, 2021, from http://www.ercot.com/

630337 CFD Based Study to Predict Rollover in an Industrial LNG Storage Tank

Abstract:

Natural Gas is transported in the liquefied form under cryogenic conditions, which makes LNG easier to transport and safer to store. LNG is stored in a highly insulated tank, despite which heat ingress from the environment is inevitable. Consequently, the lightest component (methane) evaporates. This results in a vapor stream known as Boil-off-gas (BOG) and increases the density of the remaining LNG. In an LNG terminal, when cargo is unloaded to a storage tank containing LNG of different densities, stratified layers with different liquid densities are formed. These layers further mix due to density equilibration and interface disruption under the effect of static pressure and cross the saturation temperature. This results in a large amount of BOG generation instantly, which leads to high tank pressure and a hazardous situation. This phenomenon of overall mixing and sudden generation of BOG is known as rollover. Over 24 rollover incidents have been reported over the last 60 years (GIIGNL, 2015), including those in La Spezia (Italy, 1971) and Partington (UK, 1993). In order to prevent the significant losses that arise during a rollover, it is important to develop a fundamental understanding of the boil-off-gas generation and the rollover phenomena, which is the focus of this paper.

The rollover phenomenon has been studied both theoretically as well as experimentally over the years. Germeles et al. (1975) proposed a double-diffusion LNG model by approximating methane & other components as a salt-sugar solution (thermohaline correlation) without considering boiling mechanism and turbulence during a rollover event. Heestand et al. (1983) improved this model by considering turbulence convection for heat and mass transfer between the stratified layers and rejected the salt-sugar solution model. Koyama et al. (2007), used a 3D CFD model to study rollover by considering light LNG filling only through the bottom of the tank without including the miscibility and the vapor mass removal from the liquid phase due to evaporation. Li et al. (2015) considered a 2D CFD model and induced stratification artificially to study rollover without considering the heat ingress. Hubert et al. (2019) improved the Li model by considering 3D CFD simulation with heat ingress through tank wall but did not include LNG filling mode of operation. Saleem et al. (2020) considered miscibility factors in the 2D CFD model and formulated stratification by different modes of tank filling. The major shortcoming of the model was to study the rollover under 2D tank without considering the effect of LNG unloading mode over rollover and BOG generation. These studies indicate that a 3D model of the tank that considers the effect of loading and unloading operations can represent the rollover in a more realistic manner. In the existing literature, rollover scenarios have been studied by inducing stratification artificially at the top and bottom of the tank, but the effect of unloading operation over rollover has not been studied. Since this is the most common occurrence in industry and the one of most concern to the LNG terminal operations personnel, we seek to fill these gaps in this work.

In this paper, we develop a CFD model to gain a fundamental understanding of the rollover phenomenon and the associated BOG generation. In this model, a 3D tank is used to store LNG at the receiving terminal. The tank is initialized to hold some initial level of LNG (inventory), while LNG of varying densities are allowed to fill either through the top or bottom of the tank to induce stratification. We consider heat ingress continuously through the walls of the tank, which increases the temperature of the stored LNG under the effect of local pressure (both tank top pressure and static pressure of LNG). This local pressure is used to calculate the bubble and dew point of the LNG mixture. Our model offers the flexibility to consider the LNG in the storage tank to be considered either as miscible or immiscible as necessary. In this paper, we report results from a case study to study rollover during LNG unloading operation. While this is a common operation in industry and the most susceptible to inducing rollover, it has not been previously studied in literature. Stratification is induced by filling a heavy density LNG from the bottom of the tank. Sensitivity studies on various important parameters tank size, inventory, rate of heat ingress, and density difference of LNG are also performed. Results from scaled-down studies (smaller size tank, shorter duration) and from industrial scales will both be presented. The operational benefits to the terminal from the accurate estimates of BOG generation rate and rollover time will also be discussed.

References:

1. GIIGNL Technical Study Group, 2015, Rollover in LNG Storage Tanks; Technical report.
2. Germeles, A.E., 1975. A model for LNG tank rollover 11.
3. Heestand, J., Shipman, C.W., Meader, J.W., 1983. A predictive model for rollover in stratified LNG tanks. AIChE J. 29, 199–207. https://doi.org/10.1002/aic.690290205
4. Koyama, K., 2007. CFD Simulation on LNG Storage Tank to Improve Safety and Reduce Cost, in: Koyamada, K., Tamura, S., Ono, O. (Eds.), Systems Modeling and Simulation. Springer Japan, Tokyo, pp. 39–43. https://doi.org/10.1007/978-4-431-49022-7_8
5. Li, Y., Li, Z., Wang, W., 2015. Simulating on rollover phenomenon in LNG storage tanks and determination of the rollover threshold. Journal of Loss Prevention in the Process Industries 37, 132–142. https://doi.org/10.1016/j.jlp.2015.07.007
6. Hubert, A., Dembele, S., Denissenko, P., Wen, J., 2019. Predicting Liquefied Natural Gas (LNG) rollovers using Computational Fluid Dynamics. Journal of Loss Prevention in the Process Industries 62, 103922. https://doi.org/10.1016/j.jlp.2019.103922
7. Saleem, A., Farooq, S., Karimi, I.A., Banerjee, R., 2020. CFD Analysis of Stratification and Rollover Phenomena in an Industrial-Scale LNG Storage Tank. Ind. Eng. Chem. Res. 59, 14126–14144. https://doi.org/10.1021/acs.iecr.0c02546

623091 Digital Twin of Renewable Energy-Linked Power-to-Gas (P2G) Systems: Model Building and Continuous Update for Hydrogen Producing Electrolysis

Abstract:

Solar energy and wind energy, which are widely used renewable energies, are highly volatile and intermittent, and long-term large-capacity electricity storage is required to maintain the stability of the power grid. Power-to-Gas (P2G) technology is used to generate and store gas from surplus electricity produced from renewable energy. Most P2G systems use electrolysis to produce hydrogen: the produced hydrogen is stored in a pressure tank and can be converted back to electricity through a fuel cell or hydrogen combustion engine when needed.

Hydrogen has a wide flammable range and low ignition energy compared to existing main energy sources such as gasoline or natural gas, so it has a high risk of fire and explosion, and is a material with high risk of emitting high combustion heat in case of explosion. Therefore, it is essential to evaluate the risk of hydrogen processing and storage equipment. The stability of the P2G system is an important factor in system operation along with safety. In particular, when the introduction rate of renewable energy increases, it is essential to operate a stable P2G system in order to process more excess energy.

This study aims to build a digital twin for a 500kW alkaline water electrolysis system to be built in the Saemangeum Renewable Energy National Demonstration Research Complex in Korea. We improve the digital twin model of the water electrolysis system using real-time sensing data enhanced by the real-time continuous model update. Though the current digital twin technology is focused on design and operation, we expand the applicability of digital twin by implementing functionalities for safe operation and safety management, including dynamic risk assessment, monitoring and fault detection.

In this study, based on the first principle-based models of alkaline water electrolysis cells, a digital twin of unit cells is developed, and a digital twin for the entire water electrolysis system consisting of a stack of these cells is developed. The core models consisting of the digital twin are as follows: 1) thermodynamic model for calculating Gibbs energy related to the reversible voltage, 2) electrochemical model for calculating the cell voltage according to the temperature, pressure and current, 3) electrolysis temperature calculation model, and 4) hydrogen production performance degradation model. The system/physics and failure/degradation models are composed of the above models, and operation data are combined to tune parameters of models in real-time, which establishes a digital twin of the alkaline water electrolysis system.

The developed digital twin is validated for the entire P2G system along with each of the cell units, alkaline water electrolysis system, and hydrogen compression and storage system. As a result, the developed digital twin is differentiated from the existing operation support system as a unified model for optimum control and scheduling, monitoring and fault detection, and dynamic risk assessment of the electrolysis system. When electricity production changes due to weather fluctuations, the development system enables quick start-up and system stabilization based on more accurate prediction and simulation, and dynamic risk management through prediction of functional degradation and damage of the entire system over time. It will be expected to enable real-time optimal operation and accident prevention of the alkaline water electrolysis system across the entire lifespan.

634115 Keynote Talk - Artificial Intelligence and the Modern Workforce

Abstract:

Owing to recent advances in Machine Learning and Artificial intelligence, companies across a wide range of industries are in a race to accelerate their digital transformation journey and become a data-driven enterprise where every decision and process becomes more effective and efficient, completed with higher levels of confidence, and yielding optimal results for company and customer alike. What does a successful employee of the future need to know? What will be the key characteristics of the most successful corporations? Why is it important to be a leader and not a laggard in this race? In this keynote, we will present real world examples from some of our customer engagements involving Industry 4.0 transformation. We'll highlight examples of large scale, industrial AI implementations and share key lessons learned as a preview of what's not only possible, but likely, in the future.

631978 Keynote Talk - Industry 4.0: Continuous Pharmaceutical Manufacturing Process

Abstract:

Currently, Industry 4.0 concepts are being applied to pharma industry to achieve Pharma 4.0 paradigm. Pharma 4.0 reduces the time and resources needed for continuous pharmaceutical manufacturing and also improves the product quality and production consistency. It has many advantages but also have bigger challenges on the applications of artificial intelligence (AI)/machine learning (ML), optimization, advanced process control and cyber-physical security side because of the different levels of complexities involved [1-2]. The predictive capabilities and the quality of the pharmaceutical products can be improved significantly via employing the artificial intelligence and the advanced model predictive control (MPC) system if an appropriate cyber-physical security defense is in place.

In this work, four components of industry 4.0 namely artificial intelligence (AI)/machine learning (ML), optimization, advanced control, and cyber-physical security, have been developed and implemented into the continuous pharmaceutical manufacturing process.

Four machine learning (ML) models have been trained to predict the response of continuous pharmaceutical manufacturing process and the performance of these ML models has been compared. The investigated ML methods are long short term memory (LSTM), 1D convolution neural network (CNN), random forest (RF), and artificial neural network (ANN). The best performing ML model is then implemented into the continuous pharmaceutical tablet manufacturing process for real time prediction.

A systematic framework including the methods and tools has been developed for dynamic optimization of the feeder refill strategies. The deviation of the outlet mass flow of the feeder from the targeted flow rate has been minimized to obtain the optimum value of the feeder refill parameters. The material properties also affect the refill strategy meaning that the feeder refill strategy need to be frequently optimized if there are any changes in the materials and plant. Therefore, the developed feeder model and dynamic optimization tool can save the time and recourses as well as can improve the product quality significantly.

An advanced model predictive control (MPC) system has been implemented in the continuous pharmaceutical manufacturing (CPM) pilot-plant. The CPP's and CQA's are controlled in real time. The critical control variables that have been controlled using model predictive control (MPC) system are drug concertation, powder level before tablet press, main and pre compression forces, tablet weight and hardness. A novel control strategy for powder level control in a chute placed in between blender and tablet press unit operation of continuous tablet manufacturing process has been developed, implemented and evaluated.

A systematic framework including the methods and tools have been also developed for proactive identification and mitigation of potential cyber-physical attack risk on continuous pharmaceutical manufacturing plant [2]. The cyber-physical security relevant software tools such as Snap 7, Wireshark, and Tripwire have been applied to CPM. A novel software tool named CPS (Cyber-Physical Security) has been developed for cyber-physical security of the continuous pharmaceutical manufacturing. The integrated commercially available and developed (in house) cyber-physical security tools have added an extra layer of security of our continuous pharmaceutical manufacturing pilot-plant for any unexpected attacks.

All the relevant data generated during continuous manufacturing has been systematically collected, stored and organized in a data hub (OSI PI) and cloud system as per industry 4.0 standard.

References

[1]. Bhaskar, A., Barros, F. N., Singh, R. (2017). Development and implementation of an advanced model predictive control system into continuous pharmaceutical tablet compaction process. International Journal of Pharmaceutics, 534 (1-2), 159-178.

[2]. Singh, R. (2020). The cyber-physical security of pharmaceutical manufacturing processes. Pharma., Issue 38, 53-57.

623634 Adaptive Strategies for Updating Unit Operation Models and in-Line Monitoring of Blend Uniformity in Continuous Pharmaceutical Manufacturing Process

Abstract:

Guided by U.S. Food and Drug Administration (FDA)'s Quality-by-Design (QbD) initiative[1,2] and market incentives[3], the pharmaceutical industry has heavily invested in the transition of tablet production from a traditional batch operating mode to continuous pharmaceutical manufacturing (CPM) over the past decade. The benefits of employing CPM, including but not limited to the smaller footprint, reduced human impacts, increased controllability, and shortened release time, have been clearly identified[4]. To better understand this emerging process and perform quality assessments, different mathematical models for unit operations and in-line process analytical technology (PAT) tools have been developed[4,5]. On the unit operation level, depending on the availability of process knowledge and the extent of data incorporation, these models can be classified as: (1) mechanistic models that strongly rely on process understanding; (2) data-driven models which depend only on the process data; and (3) hybrid models that integrate the first two categories[6,7]. In terms of the in-line PAT tools which monitor product quality, the performance relies heavily on data-driven global calibration models, such as partial least squares (PLS) regression[8].

The development and implementation of data-driven and hybrid models have a well-known challenge related to model maintenance. Since these models are typically trained based on historical data, they can only reflect the system for the range of process conditions, material properties, and environmental variables present at the time of model building[9]. However, shifts in the CPM process can happen over time due to changes such as environmental conditions, equipment wear and tear, lot-specific material properties, operating conditions, etc. These changes can happen suddenly, gradually, incrementally, or in a pattern[10], impacting performance of unit operations and ultimately, the critical quality attributes (CQAs) of a product. It is noted that these changes are usually not captured in typical training datasets, leading to deteriorating model prediction accuracy[10]. Therefore, appropriate model maintenance strategies need to be established. For unit operation models that involve data-driven components, model maintenance can ensure accurate representation of the manufacturing process and allow dynamic system analyses. For PAT-based chemometric models, maintenance is important for the calibration models to be robust and transferable[11,12], but traditional maintenance strategies are generally complex, costly, and time-consuming[8,11,13]. For example, depending on the diagnostics of a reduced model performance, more calibration data may need to be added, requiring additional experiments to be performed, consuming time and financial resources.

In this work, adaptive modeling strategies are developed and assessed, to serve as a model maintenance tool to cope with non-linear time-varying changes in process characteristics and/or operating conditions. The developed hybrid unit operation models for the continuous direct compaction (CDC) line[4,14] are used as baseline models. Adaptive windowing[15] and ensembled adaptive windowing[15] are selected as adaptive algorithms to be embedded into the data-driven components of the base cases. In addition to using inputs at specific time points for prediction, these algorithms incorporate change detection mechanisms, which monitor the process data streams to identify any process changes. Given such information, the algorithms can then update the data-driven components of the models. The prediction performances of the resulting adaptive models are compared against the base models in cases involving changes in process conditions, material properties, and design parameters. In terms of the PAT-based chemometric models, performance of locally weighted PLS (LW-PLS)[16], moving window[15], and recursive PLS (RPLS)[17] are evaluated in comparison to regular PLS for in-line monitoring of low dose blends in the feed frame of a tablet press using near infra-red (NIR) spectroscopy. Performance assessment involves a cross-prediction between datasets comprising of NIR spectra collected at different feed frame speeds with physical baseline differences in the NIR spectra. Cross-prediction can help to evaluate the effectiveness of adaptive methods to prevent the anticipated deterioration of prediction accuracy. Case studies involving time-varying process condition changes are also developed for comparison. The results of adaptive unit operation and chemometric models demonstrate a more accurate prediction performance in systems with changes in operating conditions. Using these adaptive modeling strategies, the models can be updated in real-time or periodically to enhance the predictability of CQAs throughout the process development and operation phases[18].

References

(1) Yu LX, Akseli I, Allen B, et al. Advancing Product Quality: a Summary of the Second FDA/PQRI Conference. *AAPS J.* 2016;18(2):528-543.

(2) Yu LX, Kopcha M. The future of pharmaceutical quality and the path to get there. *Int J Pharm.* 2017;528(1-2):354-359.

(3) Boukouvala F, Niotis V, Ramachandran R, Muzzio FJ, Ierapetritou MG. An integrated approach for dynamic flowsheet modeling and sensitivity analysis of a continuous tablet manufacturing process. *Comput Chem Eng.* 2012;42:30-47.

(4) Wang Z, Escotet-Espinoza MS, Ierapetritou M. Process analysis and optimization of continuous pharmaceutical manufacturing using flowsheet models. *Comput Chem Eng.* 2017;107:77-91.

(5) Ierapetritou M, Muzzio F, Reklaitis G. Perspectives on the continuous manufacturing of powder-based pharmaceutical processes. *AIChE Journal.* 2016;62(6):1846-1862.

(6) Zendehboudi S, Rezaei N, Lohi A. Applications of hybrid models in chemical, petroleum, and energy systems: A systematic review. *Applied Energy.* 2018;228:2539-2566.

(7) von Stosch M, Oliveira R, Peres J, Feyo de Azevedo S. Hybrid semi-parametric modeling in process systems engineering: Past, present and future. *Comput Chem Eng.* 2014;60:86-101.

(8) Wise BM, Roginski RT. A Calibration Model Maintenance Roadmap. *IFAC-PapersOnLine.* 2015;48(8):260-265.

(9) Webb GI, Hyde R, Cao H, Nguyen HL, Petitjean F. Characterizing concept drift. *Data Mining and Knowledge Discovery.* 2016;30(4):964-994.

(10) Gama J, Žliobaitė I, Bifet A, Pechenizkiy M, Bouchachia A. A survey on concept drift adaptation. *ACM Computing Surveys.* 2014;46(4):1-37.

(11) Zeaiter M, Roger JM, Bellon-Maurel V, Rutledge DN. Robustness of models developed by multivariate calibration. Part I. *TrAC Trends in Analytical Chemistry.* 2004;23(2):157-170.

(12) Besseling R, Damen M, Tran T, et al. An efficient, maintenance free and approved method for spectroscopic control and monitoring of blend uniformity: The moving F-test. *J Pharm Biomed Anal.* 2015;114:471-481.

(13) Zhong L, Gao L, Li L, Zang H. Trends-process analytical technology in solid oral dosage manufacturing. *Eur J Pharm Biopharm.* 2020;153:187-199.

(14) Chen Y, Ierapetritou M. A framework of hybrid model development with identification of plant-model mismatch. *AIChE Journal.* 2020.

(15) Kadlec P, Grbić R, Gabrys B. Review of adaptation mechanisms for data-driven soft sensors. *Comput Chem Eng.* 2011;35(1):1-24.

(16) Hazama K, Kano M. Covariance-based locally weighted partial least squares for high-performance adaptive modeling. *Chemometrics and Intelligent Laboratory Systems.* 2015;146:55-62.

(17) Joe Qin S. Recursive PLS algorithms for adaptive data modeling. *Comput Chem Eng.* 1998;22(4-5):503-514.

(18) Gama J, Medas P, Castillo G, Rodrigues P. Learning with Drift Detection. 2004; Berlin, Heidelberg.

626250 Data-Driven Approaches Towards Equipment Health-Classification and Predictive Monitoring in Drug Product Manufacturing

Abstract:

Digitalization is the central pillar towards the introduction of Industry 4.0 in pharmaceutical manufacturing. However, sophisticated data-driven approaches are yet to be explored to fully take advantage of the abundance of recorded process data available. Drug manufacturing is a highly regulated field. Thus, regulators require manufacturers to store process data over many years for backtracking purposes. This work explores the potential to use such data for predictive maintenance purposes.

This work presents a data-driven approach for equipment monitoring and the implementation of equipment classification and predictive maintenance in an aseptic filling line in drug product manufacturing. The goal is to leverage stored short-term process monitoring data for a secondary application in long-term equipment monitoring. Multiple years' worth of industrial process data are used, which include process monitoring data and logbook information regarding each process run. Process monitoring data is mainly acquired through Supervisory Control and Data Acquisition (SCADA). Monitored variables include pressure, temperature, or process step duration. Furthermore, logbook information records details regarding equipment and process failures as well as maintenance data including measures taken and maintenance intervals.

In a first step, process data was pre-processed and different classification algorithms were applied. Thereby, equipment failures were shown to be clearly linked with process monitoring data through, either, observable shifts in principal curves or peaks representing events. Short-term noise in the process data through variation caused by manual operation and batch-wise production were successfully minimized and long-term based effects due to equipment deterioration were identified. An analysis of results obtained from different classification algorithms for equipment monitoring is presented.

In a next step, different machine learning algorithms are applied to the process data and compared in their ability to predict equipment deterioration and failure. Additionally, optimized maintenance strategies for equipment are formulated. Finally, a decision-support framework is being developed, where expert knowledge will be included for defining counter measures taken in the case of alarming changes in equipment status.

630095 An Equation-Oriented Model of a Trickle-Bed Reactor for Hydrodesulfurization Process Analysis and Digital Twin Applications

Abstract:

The Fourth Industrial Revolution is marked by the convergence of digital, physical, and biological technologies, bringing relevant tools to allow deep process analysis and optimizations, thus enabling more assertive decision-making. One such tool directly linked to rigorous modeling and simulation is the process digital twin. When a digital twin is connected to powerful data sources (i.e., rigorous thermodynamic basis, industrial data sets, and detailed phenomenological models), it can be used for process design [1,2]. In this sense, the industrial sector, particularly the oil and gas industry, can benefit of digital twin approaches to improve process efficiency. The Hydrodesulfurization (HDS) process is considered a well-known and established technology used in refineries to remove sulfur compounds, such as sulfides and disulfides, from petroleum fractions. HDS consists of a catalytic process capable of treating a contaminated oil charge stream through a hydrogenation reaction under controlled operating conditions that are highly dependent on the oil physicochemical characteristics [3]. However, the definition of such operating conditions, as well as the evaluations related to reactant characteristics and catalysts need special attention.

In this work, the improvement of a rigorous HDS process model, based on a trickle-bed reactor block [4] in AVEVA Process Simulation is presented. The previously validated phenomenological model [5,6] has been connected to the AVEVA thermodynamics database for comprehensive process analysis. The resulting simulation enables the description of the sulfur removal and the formation of H_2S process throughout the reactor, in addition to satisfactorily predicting the effects on the conversion of the organic sulfur compounds through changes in nominal trickle-bed reactor conditions. The obtained relationship employing this model between the gas/oil ratio and the partial pressure of hydrogen sulfide presented a realistic behavior, enabling improvements on sulfur removal. Operating conditions, such as temperature and pressure, present consistent performance associated with effectiveness of desulfurization. The concentrations of the compounds in liquid phase and on the catalyst surface are analyzed throughout the reactor confirming experimental and industrial results [5,6].

References

[1] Macchi, M., Roda, I., Negri, E., & Fumagalli, L. (2018). Exploring the Role of Digital Twin for Asset Lifecycle Management. IFAC-PapersOnLine, 51(11), 790–795.

[2] Rosen, R.; Von Wichert, G.; Lo, G., & Bettenhausem, K. D. (2015). About The Importance of Autonomy and Digital Twins for the Future of Manufacturing. In IFAC-PapersOnLine, 48:567–72. Elsevier Ltd.

[3] Mendez, F.J.; Franco-Lopez, O.E.;. Bokhimi, X.; Solis-Casados, D. A.; Escobar-Alarcon, L., & Klimova, T. E. (2017). Dibenzothiophene hydrodesulfurization with NiMo and CoMo catalysts supported on niobium-modified MCM-41, Appl. Catal. B Environ. 219, pp 479–491.

[4] Barros, A.; Bishop, B..; Tavernard, A.; Lima, F. V., & Bispo, H. (2020). Development of an Equation Oriented Model Applied to a Hydrodesulfurization (HDS) Process. AIChE Annual Meeting.

[5] Neto, A. T. P., Fernandes, T. C. R. L., da Silva Junior, H. B., de Araújo, A. C. B., & Alves, J. J. N. (2020). Three-phase trickle-bed reactor model for industrial hydrotreating processes: CFD and experimental verification. Fuel Processing Technology, 208.

[6] Korsten, H., & Hoffmann, U. (1996). Three phase reactor model for hydrotreating in pilot trickle bed reactor, AIChE Journal, 42, pp 1350-1360.

634116 Keynote Talk - Digital Transformation, Data Science, and Industrial Intelligence

Abstract:

Data analysis has played an important role in process monitoring and operations for over three decades, with early methods focusing on multivariate statistics and artificial neural networks. Over the last ten years, along with the rise of big data, industrial internet of things (IIoT), statistical machine learning, and deep neural networks, digitalization and data analytics have taken place in nearly all industry sectors with an unprecedented pace. The McKinsey report in 2011 envisioned that analyzing large data sets would become a key basis of competition, underpinning new waves of productivity growth and innovations. Recent surveys by McKinsey revealed that the adoption of machine learning varied significantly from sector to sector.

In this keynote we offer a few perspectives on the development of data analytics and machine learning in process systems engineering and how we can maneuver their healthy development to harness the full potential of the digital revolution. Digitalization and the IIoT adoption will provide the process industries with massive amount of data from process operations to process maintenance. The power of data science and analytics from statistical learning methods to deep neural networks will need to be scrutinized and tailored for process applications. In driving toward industrial intelligence, we attempt to address the following big-picture questions in this talk.

i) What is the role of data science and analytics in transforming data into information and knowledge?

ii) Will human be out of the loop and replaced by artificial intelligence or autonomous systems?

iii) Will industry need deep learning or explainable machine learning?

iv) Will first principles be replaced with data-driven deep learning?

634131 Keynote Talk - Decision Making in Digitalized Value Chain Optimization

Abstract:

With digitalization being such a hot topic in control and real-time optimization, this talk focuses more narrowly on tactical and strategic decision making which typically has more humans in the loop. After briefly introducing ExxonMobil's Downstream value chain, I will highlight two broad themes: Modeling competition between players and tractable optimization under uncertainty. In the first theme, I will introduce problems in which multiple players vie with one another to maximize individual profit in a non-cooperative competitive market. This competitive setting is interesting and worthy of study because the majority of prevailing process systems engineering models largely overlook non-cooperative strategies that exist in real-world markets. In the second theme, I will discuss the decision rule approach to optimization under uncertainty and how we have applied it to several PSE problems. I will conclude with some future research topics that I hope are of interest to the broad community.

624181 Predicting the Diameters of Droplets Produced in Turbulent Liquid-Liquid Dispersion

Abstract:

Liquid-liquid dispersion is a common chemical manufacturing batch process typically performed in agitated tanks. The physics informing the droplet size distribution is a complex convolution of impeller speed, impeller type, fluid properties, and flow conditions. In this work, we present three *a priori* mechanistic modeling approaches for predicting the droplet diameter distributions as a function of system operating conditions. In the first approach, called the two-fluid approach, we use high-resolution solutions to the Navier-Stokes equations to directly model the flow of each phase and the corresponding droplet breakup/coalescence events. In the second approach, based on an Eulerian-Lagrangian model, we describe the dispersed fluid as individual spheres undergoing ongoing breakup and coalescence events per user-defined interaction kernels. In the third approach, called the Eulerian-Parcel model, we model a sub-set of the droplets in the Eulerian-Lagrangian model to estimate the overall behavior of the entire droplet population. We discuss output from each model within the context of predictions from first principles turbulent theory and measured data. We then discuss the scalability of each method to various operating volumes/conditions.

624622 Computational Fluid Dynamics Study to Develop a Scale-up Strategy for Non-Newtonian Fluids in Pharmaceutical Industry

Abstract:

The generic pharmaceutical industry relies on the speed and rigor in the drug development process. Getting it right the first time requires detailed product and process understanding. Various unit operations like sterilization, milling, mixing, etc. govern the rapidity of the process. Mixing is one such operation, developing a proper technical understanding can yield highly productive results and can reduce experimental work thereby proving to be cost and time effective. This work provides a comprehensive approach for understanding mixing in different equipment for several parameters and scale-up strategy using computational fluid dynamics.

The first question to ask is what comprises good mixing. The judging parameters include the uniform distribution of energy with less dissipation and getting the right CQAs in the samples. The major challenges in mixing can be attributed to non-uniformity and unpredictability in scale-up operations. Computational fluid dynamics (CFD) is an approach that is coming into action in the pharmaceutical industry. CFD Software (ANSYS in this example) can be leveraged to understand mixing profiles and help in analyzing the technical aspects and provide a generalized methodology for the problem.

As the complexity of molecules has increased and with process intensification, the need to develop technical solutions faster and having a standardized approach will support the entire industry to reach its goal of serving people better. This study gives a comparison of equipment based on histograms, contour plots, and vector plots for shear stress, velocity, and molecular viscosity, along with the discrete phase model to draw inferences and conclusions for non-Newtonian fluid systems. Transient simulations are used to determine the particle distribution in the mixing vessel volume being dispersed from a specified location. This helps in finding time to achieve uniformity throughout the volume. Using CFD the problem at lab scale is resolved by determining the CPPs which in turn gives us the ability to scale-up the product in a plant to achieve desired targets.

The impact of this study can be huge, as it can contribute not only to reduce experimental work at lab scale but can also improve visibility in the scale-up plant operations. It further adds as a standardized method for such issues.

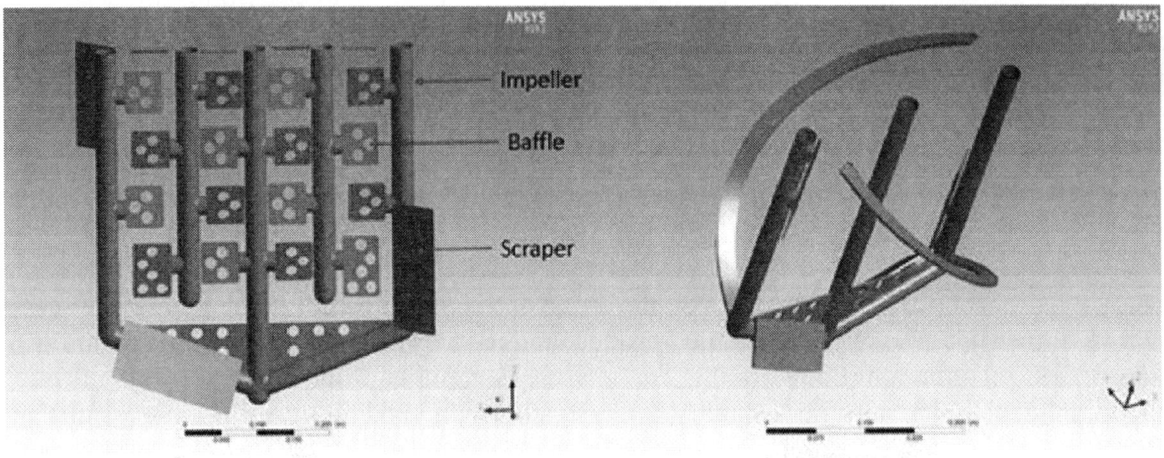

Anchor Impeller

Total Volume = 100L

Product Volume used = 85L

Helical Impeller

Figure 1: CAD models of Anchor and Helical impellers available at plant scale

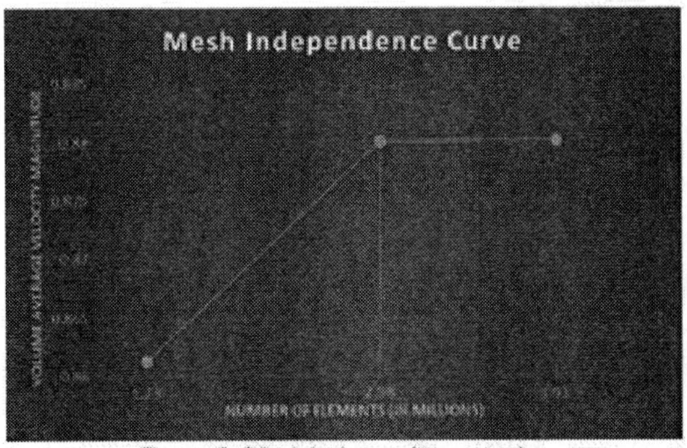

Figure 2: Mesh independence study

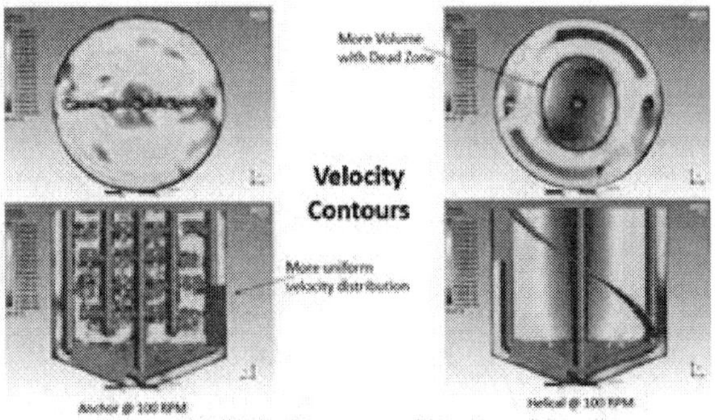

Figure 3: Velocity Contours of plant scale study

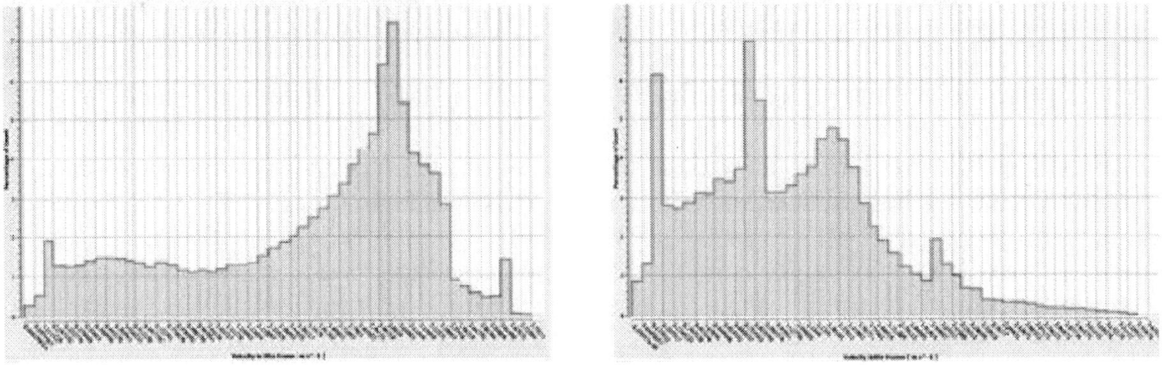

Skewed Distribution

Uniform Distribution

Figure 4: Velocity Histograms for Helical and Anchor impellers for plant scale study

627762 Modelling the Effect of Ice Nucleation Stochasticity on Batch Heterogeneity of Freezing Processes of Vials on a Shelf

Abstract:

Introduction

Freezing and freeze-drying are widely used processes to improve the stability and thus shelf life of pharmaceutical formulations. Ice nucleation is the first step of the freezing process; it is an activated process that requires cooling of the product beyond its equilibrium freezing temperature. It occurs at different times and temperatures for individual vials, even in case they experience identical cooling profiles. This variance was identified as a main source of batch heterogeneity across frozen vials on a shelf [1]. Such heterogeneity is a major risk for industrial processes involving freezing steps since several product attributes depend strongly on the freezing history of each vial: The ice crystal morphology in a vial depends both on the temperature at which ice nucleation occurs and on the time required for complete solidification after nucleation, i.e. the solidification time [2]. Finally, stochasticity of ice nucleation depends, to some extent, on the process conditions: Generally, one observes lower average nucleation temperatures and a broader distribution in dust-free GxP settings compared to the laboratory, thus rendering the transfer of a process design from laboratory to production scale challenging. To mitigate the effects of this stochasticity, process technologies have been developed that artificially induce ice nucleation at the same time in all vials, so called controlled nucleation techniques [1].

While there is experimental evidence that these techniques indeed reduce heterogeneity across the shelf, experimental studies to quantify heterogeneity are challenging due to the requirement of analyzing several product attributes in a statistically significant number of vials. Therefore, a shelf-scale model of the freezing process is required to deepen the understanding of batch heterogeneity and to optimize the process design. To the best of our knowledge, such model is not available in literature and the theoretical understanding of the relevant mechanisms leading to heterogeneity is limited. Besides the stochasticity of ice nucleation, variability in heat transfer was identified as major source of heterogeneity, in the form of spatial differences across the shelf and due to geometric imperfections of vials [4].

We thus propose a modelling framework for the freezing process of an arbitrary number of vials on a shelf to understand and identify all potential mechanisms of batch heterogeneity with a focus on the stochasticity of ice nucleation. With the help of this model, we investigate the effect of various process parameters on heterogeneity to aid process design and upscaling.

Methods

The model describes the freezing process of vials on a shelf in a mechanistic way taking into account the geometry of the system, heat transfer, and ice nucleation. Each vial is considered as thermally homogeneous while it evolves from liquid to a partially frozen, and to a completely frozen state. The governing equations account for the vial's physical state and ice nucleation is described as an inhomogeneous Poisson process [5].

While experimental studies on batch heterogeneity mostly focus on measuring product attributes, we characterize heterogeneity based on three characteristic quantities of the freezing process. These are the nucleation time, nucleation temperature and solidification time for each vial of the simulated system.

We have simulated sets of freezing experiments and generated the corresponding histograms of these quantities. To identify potential mechanisms for heterogeneity, we investigated the effect of ice nucleation parameters and their variability, heat transfer parameters and their variability and process control strategies on these distributions.

Results

We have performed simulation studies of a system comprising 7x7 vials on a shelf, a typical batch size for experimental studies in laboratory scale freeze-dryers. In addition to the direct effect of the stochasticity of ice nucleation on heterogeneity, we account also for an indirect role played by ice nucleation, namely heat transfer between vials triggered by the heat released upon nucleation. Since nucleation is stochastic, the occurrence of such thermal effects adds to the heterogeneity across the shelf. Figure 1 illustrates both these effects on the quantities characterizing the freezing process, expressed in terms of the intensity of the interaction. We use the heat transfer coefficient between vials, as parameter to quantify this interaction.

In case of thermally independent vials, i.e. $k_{int} = 0$, the distributions of the characteristic quantities correspond to the freezing process of single vials: The only source of variability between vials is the stochasticity of ice nucleation. For sufficiently strong interaction, i.e. $k_{int} > 10\ Wm^{-2}K^{-1}$, we observed a bimodal nucleation time distribution as shown in Fig. 1a. In this case, a first group of vials nucleated randomly across the shelf and delayed the nucleation events of the remaining vials. These early nucleating vials then act as "local hotspots" on the shelf that temporarily slow down the cooling process of neighboring vials. The formation of distinct populations corresponding to early and late nucleating vials can be seen in the bivariate distribution of nucleation times and solidification times in Fig. 1d. These populations differ both in nucleation and solidification times, thus creating a significant heterogeneity across the batch.

The effect of thermal interaction on the nucleation temperature distribution was smaller than for the other two quantities; we only identified a weak trend towards higher temperatures for increasing interaction as shown in Fig 1a-c. This observation, namely that in some situations the distributions of nucleation time and of temperature behave differently, is of importance for process monitoring. Commonly, the two distributions are used interchangeably; nucleation times in literature are detected with a camera and converted into temperatures based on reference temperature measurements [3]. Based on our model results, we hypothesize that such approach may be misleading and suggest the direct measurement of both quantities, e.g. via infrared thermography.

We can use these outcomes to improve the freezing process design. In case of freeze-drying processes, a homogeneous crystal morphology across the shelf is required. The crystal morphology correlates directly with the solidification time, so that a narrow distribution of the solidification times across the shelf is desired. Consequently, one should aim to operate the process in a regime with low thermal interaction. In practice, this may be achieved by arranging vials in a geometry with less direct contact and by arranging them directly on the shelf without tray.

In general, we can apply this modelling framework to study not only the impact of the stochasticity of ice nucleation, but all potential sources of heterogeneity. Our model can be extended with ease to simulate the additional heat transfer to edge vials, as well as a random variability of the vial-to-shelf heat transfer coefficient.

Conclusion

We present for the first time a mechanistic model for the freezing of an arbitrary number of vials on a shelf that is capable of capturing the stochasticity of ice nucleation and its effects on the freezing process. It proves to be a useful tool in improving the understanding of batch heterogeneity in freezing processes. The model enables us to assess the impact of process parameters, such as varying cooling profiles or techniques like controlled nucleation, on the freezing outcome. It also allows the simulation of the freezing process for both GxP and non-GxP settings based on the specific ice nucleation kinetics of each setting; thus aiding upscaling and technology transfer between freezing devices. The developed model thus serves as a powerful simulation tool in the design and optimization of freezing and freeze-drying.

Acknowledgement

The authors thank The Janssen Pharmaceutical Companies of Johnson & Johnson for the support in the course of this project.

References

[1] A. Arsiccio et al.: Vacuum Induced Surface Freezing as an effective method for improved inter- and intra-vial product homogeneity, *European Journal of Pharmaceutics and Biopharmaceutics* 128 (2018) 210–219.

[2] A. Arsiccio, A. A. Barresi, and R. Pisano: Prediction of Ice Crystal Size Distribution after Freezing of Pharmaceutical Solutions, *Crystal Growth & Design* (2017), 17, 4573–4581.

[3] L. C. Capozzi, R. Pisano: Looking inside the 'black box': Freezing engineering to ensure the quality of freeze-dried biopharmaceuticals, *European Journal of Pharmaceutics and Biopharmaceutics* 129 (2018) 58–65.

[4] B. Scutella et al.: How Vial Geometry Variability Influences Heat Transfer and Product Temperature During Freeze-Drying, *Journal of Pharmaceutical Sciences* 106 (2017) 770-778.

[5] G. M. Maggioni and M. Mazzotti: Modelling the stochastic behaviour of primary nucleation, *Faraday Discussions* (2015), 179, 359.

629683 API Crystallization Process Development for Improved Consistency, Bulk Properties, and to Enable Downstream Processing

Abstract:

A revised crystallization process for an Active Pharmaceutical Ingredient (API) was developed to improve the physiochemical and bulk powder properties of the isolated product. The first generation process consisted of an unseeded cooling crystallization that provided material with low bulk density, a propensity to agglomerate on handling and storage, and was not amenable to formulation. Additionally, a clarifying filtration was not possible due to solubility limitations of the API at the target operating volume. Development of the crystallization process explored the use of seed crystals generated by wet-milling and jet-milling, the use of in-situ wet milling during the crystallization, and thermalcycling to improve the bulk properties of the resultant powders. Optimization of these parameters led to a second generation process that is run at a higher operating volume to accommodate a clarifying filtration, distilled to a target concentration, seeded with jet-milled seed material, and subjected to a series of thermalcycles to ripen the product crystals. The resulting API is isolated with an increased bulk density, higher aspect ratio, and improved handling during formulation. The second generation process was studied as part of a fractional factorial Design of Experiments to assess and rank the impact of all of the processing parameters on the isolated product's quality attributes. It was determined that the physical properties of the seed crystals, the seeding point, and the temperature targets of the thermal cycling had the greatest impact on the product particle size distribution and bulk powder properties.

622378 Loss-in-Weight Control of a Low Dose Powder Feeding System

Abstract:

Accurate and robust powder feeding is the key for successful continuous pharmaceutical manufacturing. Feeding systems available on the market are typically suitable for feed rates above 200 g/h. We show the development of loss-in-weight (LIW) control for a novel feeding system capable of continuously feeding powder at feed rates in the range of few g/h. Our in-house developed prototype is equipped with a LIW scale, a powder pump and a scraping device. Additionally, a high precision gain-in-weight (GIW) scale is used as reference. First, a feed-forward (FF) control strategy was established to compensate for the material dependent displacement density along the powder container. This approach is similar to the feed factor correction based on the hopper fill level of commercially available large scale feeders. In a second step, an iterative learning (IL) control strategy was developed, which updates the model for FF control after regularly spaced time intervals based on the measured feed rate deviation from the set-point.

In the presentation, we compare operation with constant powder pump speed to feed-forward controlled experiments. It can be shown that the feed rate deviations, caused by the reproducible displacement density, can be mitigated. Furthermore, we show the principle of IL control in order to compensate model inaccuracies as a result of changing material behavior due to batch-to-batch variability or inconsistent powder conditioning by the operator.

A LIW control strategy for a novel low dose feeding system was developed, successfully transferred to the prototype and tested. Accurate and robust feeding for feed rates as low as 1.5 g/h could be achieved and by means of IL control, the effect of changing material behavior could be significantly reduced.

624021 Toxicity of Orally Administered Magnetic Imaging Nanoprobes in an Inflammatory Intestinal Epithelial Cell Model

Abstract:

SiO_2-coated superparamagnetic iron oxide nanoparticles (SPION) can be efficiently produced in a scalable, single-step flame process.[1,2] The SiO_2-layer on the particle surface facilitates subsequent chemical modifications, and the magnetic iron oxide core makes the SPIONs suitable as MRI contrast agents. We hypothesized that functionalizing SPIONs with ligands that target overexpressed biomarkers in the gastrointestinal tract (GIT) can allow non-invasive imaging of diseases such as inflammatory bowel disease (IBD) *in vivo*. Biomarkers in preclinical IBD models include ICAM1,[3] which could be targeted using functionalized SPIONs. However, exposing damaged and inflamed cells to such functionalized nanoparticles might have the potential to cause immunotoxicity and compromise their safe use as contrast agents. Pure iron oxide and SiO_2-coated iron oxide nanoparticles were made through flame spray pyrolysis.[1] Monoclonal antibody anti-ICAM1 is conjugated onto the SiO_2-coated particle surface using click chemistry to target the IBD associated biomarker ICAM1. Filter-grown Caco-2 cells are treated with a mixture of inflammatory inducing agents to resemble the inflamed state of IBD. Both the antibody-functionalized, non-functionalized and pure SPION particles are evaluated for their cytotoxic and genotoxic effects, as well as the morphological changes in the cell monolayer. Cytotoxicity and genotoxicity is examined using CellTiterGlo assay and Comet assay, respectively. Cell integrity is examined by measuring permeability of Lucifer yellow across the cell monolayer. A dose dependent correlation between SPIONs and toxicity is expected to be found. Global proteomics of inflamed Caco-2 cell model has shown the upregulation of biomarkers, such as ICAM1, in the apical plasma membrane.[3] These results provide guidance for appropriate functionalization of SPIONs to target the inflamed tissue. The inflamed cell model is likely to demonstrate a higher sensitivity to the SPIONs compared to the healthy cell monolayer. Adverse toxic effects of nanomaterial based imaging probes for use *in vivo* is important to assess. Evaluation of cytotoxic and genotoxic effects, as well as changes in permeability of inflamed Caco-2 cell monolayers provide an estimation of the safety to use functionalized SPION as a diagnostic tool *in vivo*.

References: [1] Teleki, A., M. Suter, P.R. Kidambi, O. Ergeneman, F. Krumeich, B.J. Nelson, and S.E. Pratsinis, *Chem. Mater.* **21**, 10, (2009). [2] Teleki, A., F.L. Haufe, A.M. Hirt, S.E. Pratsinis, and G.A. Sotiriou, *RSC Adv.* **6**, 26, (2016). [3] Asad, S., Wegler, C., D. Ahl, C.A.S. Bergström, M. Phillipson, P. Artursson and A. Teleki, *J. Pharm. Sci.* **110**, 1 (2021).

627347 3D Reporter Gene Assay for High-Throughput Drug Screening

Abstract:

Traditional cell-based assay has high probability of failure to identify effective and safe compounds in early drug discovery. Approaches to address this problem include engineering of cancer cell lines to indicate the drug mechanism of action, and applying of 3D cancer models to simulate the cancer tissue physiology. In this research, cancer cell lines MCF7, Panc-1 and A549 are engineered as reporter cells and are cultured on polyethylene terephthalate (PET)-based 3D scaffold for high throughput drug screening. The assay reports the regulation of two prominent cancer biomarkers, survivin and human telomerase reverse transcriptase (hTERT), which are silenced in normal adult cells but are overexpressed in more than 85% tumor tissues. Fluorescent proteins as the reporter were used to achieve a non-invasive and real-time monitoring of biomarker expression. Fluorescence intensity reduction is observed after testing the biomarker inhibitors, such as YM155 and doxorubicin for survivin, and several natural products for hTERT. This assay offers both molecular and physiological predictions of tissue-specific responses to antitumor agents, thus facilitating early-stage cancer drug discovery.

629697 Developing Fluorescent Viral Surrogate Particles As an Alternative to Live Virus in Viral Clearance Validation

Abstract:

When manufacturing protein-based biologic drugs in mammalian hosts, viral contaminations are of particular concern as they put patients at risk of contracting potentially fatal illnesses and cause manufacturing disruptions, pressing companies into costly decontamination and recall actions. Given these serious repercussions, regulatory agencies have issued guidelines requiring proof of sufficient viral clearance across the downstream process in case a contamination event should occur. This proof of clearance is currently accomplished by performing viral spiking studies. Viral spiking studies are executed by introducing high concentrations of live virus into drug-containing load materials and clearance capacity is evaluated across each downstream unit operation. These studies are costly in both time and money as they are typically performed externally with a certified contract research organization (CRO) with biosafety level (BSL)-2 capability; manufacturers will not knowingly introduce contaminating virus into their processes on-site owing to fears of recurring contamination. In this work, we address this technical gap by developing a series of fluorescent viral surrogate (FVS) particles, which are easily quantifiable through fluorescence-based measurements and have well-defined physicochemical characteristics aligning with enveloped and non-enveloped model viruses currently employed in viral clearance validation. Non-enveloped FVS's were synthesized by covalently modifying Fluospheres™ carboxylate-functionalized particles with fluorescent-labeled model proteins. Bare and coated Fluospheres™ were characterized based on size, charge, and relative hydrophobicity, and these data were compared to those of live virus. To quantify FVS's, a dead-end flow filtration unit operation was optimized for capture of FVS's from solution to achieve limits of detection comparable to $TCID_{50}$ and plaque titer assays (used to quantify live virus). These optimized FVS's will be characterized using clearance mechanisms across various separation unit operations, such as viral filtration (VF), affinity (AC), ion exchange (IEX), and hydrophobic interaction (HIC) chromatography. This will be done to formulate structure-function relationships and to assess the feasibility in employing FVS technology as an alternative to live virus in viral clearance validation.

632933 Pro-D-Twin: Practitioner Friendly Platform for Process Development, Visibility and Control

Abstract:

Expediting the product/ process development and tech transfer efforts and getting the product right the first-time, demand adequate understanding of the product and process. Institutionalizing such an approach in product development requires empowering of the practitioners in their decision-making steps, with friendly tools to leverage appropriate technically grounded capabilities linking process inputs to outputs. Such capabilities can be in the form of either simple calculations or complex simulations providing different levels of insights. Ease of use of such capabilities and the ability to provide required guidance in almost real time are key for effective adoption by the practitioners, and to realize the benefits. This poster focuses on how practitioners can leverage ProDTwin platform (www.prodtwin.com) for routine process needs.

623001 Extractionscore: A Quantitative Framework for Evaluating Synthetic Routes on Predicted Liquid-Liquid Extraction Performance

Abstract:

A multitude of metrics exist to assign scores to synthetic routes within computer-aided synthesis planning (CASP) tools. A quantitative scoring method is necessary to identify the most promising synthetic approaches to a molecule. Popular metrics include reagent cost, [1] quantifications of synthetic complexity, [2] and confidence in predicted reaction outcomes and diversity of reactions employed. [3] However, relatively little attention in the CASP field has been paid to the impact of side products on separation sequences, in part due to a lack of experimental data.

To address this issue, we have developed a method of quantifying liquid-liquid extraction performance for synthetic routes, which we call ExtractionScore. ExtractionScore uses existing reaction outcome predictors to predict side products in a reaction mixture, and chemical property prediction to estimate partition coefficients. Separation factors for impurities in the reaction mixture are then calculated and used to assign a numeric score for the reaction. The ExtractionScore for a synthetic route is calculated from the scores of each reaction in the route.

We have evaluated ExtractionScore by comparing the industrially practiced synthesis routes to 200 active pharmaceutical ingredients and precursors thereof with routes suggested by the IBM RXN retrosynthesis planner. In 72% of cases, the industrially practiced route scores higher on ExtractionScore than the IBM RXN suggested routes, indicating ExtractionScore's potential usefulness as a quantification of industrially relevant features of a synthesis route.

1. Szymkuć, ; Gajewska, E. P.; Klucznik, T.; Molga, K.; Dittwald, P.; Startek, M.; Bajczyk, M.; Grzybowski, B. A. Computer-assisted synthetic planning: The end of the beginning. *Angewandte Chemie International Edition* **2016**, *55*, 5904–5937.
2. Coley, C. W.; Rogers, L.; Green, W. H.; Jensen, K. F. SCScore: Synthetic complexity learned from a reaction corpus. *Journal of Chemical Information and Modeling* **2018**, *58*, 252–261.
3. Schwaller, ; Petraglia, R.; Zullo, V.; Nair, V. H.; Haeuselmann, R. A.; Pisoni, R.; Bekas, C.; Iuliano, A.; Laino, T. Predicting retrosynthetic pathways using transformer-based models and a hyper-graph exploration strategy. *Chemical Science* **2020**, *11*, 3316–3325.

623171 Accelerating Antibody Discovery with Cell-Free Systems

Abstract:

The COVID-19 pandemic has highlighted the need for rapid, high-throughput, and robust therapeutic antibody screening platforms. Advances in both synthetic selections and antibody repertoire mining have enabled the rapid identification of hundreds of candidate binders in a single experiment. However, the expression and evaluation of these candidate antibodies still remains a major bottleneck in the discovery pipeline due to labor intensive steps and process throughput mismatches. In this work, we present a newly developed workflow that leverages cell-free protein synthesis (CFPS) and an Echo® 525 acoustic liquid handler to enable the expression and evaluation of hundreds of antibodies in a single experiment. Our workflow consists of a cell-free DNA assembly step to generate antibody expression templates, a modified *E. coli* CFPS system to express antibodies, and the AlphaLISA protein-protein interaction assay to evaluate binding. Each step of this workflow can be carried out entirely within 384-well plates without purification steps and without ever going into cells, making it easily end-to-end automatable. The resulting platform is capable of evaluating more than an order of magnitude more candidates in less than one third of the time when compared to state-of-the-art antibody discovery pipelines. To validate the platform, we express a diverse set of 122 published COVID-19 neutralizing antibody candidates in triplicate in a single experiment. We evaluated these antibodies by mapping their target epitopes and evaluating their ability to compete with ACE2, the SARS-CoV-2 target human receptor. Our results are largely consistent with published literature on the antibodies, and we were able to easily distinguish the best reported candidates based on our data. We expect that these advancements will accelerate antibody discovery and development not only for viral diseases like COVID-19, but for the entire antibody industry.

630002 A Multiple Molecular Endpoints HT Screening of Embryotoxic Chemicals Based on ESC-EGFP with Human Gene Promoters Associated with Specific Embryonic Development Pathways

Abstract:

Currently, there are thousands of chemicals in common use; however, only a small number of them have sufficient toxicity evaluation, and even less information for the specific embryo-related toxicity field: embryotoxicity and developmental toxicity. Embryonic stem cell test (EST) is the only accepted *in vitro* method for assessing embryotoxicity without animal sacrifice. However, the implementation and application of conventional EST for regulatory embryotoxicity screening are impeded by its labor intensity (technical complexity), long testing period (low throughput), and poor robustness (limited differentiation lineage endpoints) for the screening of thousands of drugs and chemicals in use or under development.

A key step in the understanding of embryonic development is to construct a systemic and comprehensive gene regulation network involved in the development-related pathways. In our research, a novel EST with mouse embryonic stem cells (mESCs) expressing enhanced green fluorescent protein (EGFP) driven by human gene promoters, including those associated with survivin in the apoptosis, Oct4 (POU5f1) in pluripotency, and SOX17 and TUBB3 in tissue-specific differentiation pathways, were developed for use in high throughput (HT) screening of potential embryotoxic chemicals and their effects on embryonic development. Our hypothesis was that embryotoxic chemicals would down-regulate these gene promoters and thus decrease EGFP expression. mESCs expressing EGFP driven by the constitutive CMV promoter was used to quantify proliferation and cytotoxicity of tested chemicals. These EGFP-expressing mESCs were cultured in two-dimensional (2D) plate or three-dimensional (3D) fibrous scaffolds in microbioreactors on a multi-well plate, and EGFP fluorescence signals as indicative of cell responses to chemicals were monitored non-invasively using a fluorescent plate reader. The downregulation of these four gene promoters by chemicals with known embryotoxicity were first tested with strongly embryotoxic hydroxyurea and retinoic acid, weakly embryotoxic salicylic acid sodium salt, and non-embryotoxic acrylamide and penicillin. Then, thalidomide, which has not been classified for its embryotoxicity but its past use by pregnant women had caused severe birth defect, was tested. The results from the assay showed that thalidomide down-regulated all four gene promoters, indicating its strong embryotoxicity and damaging effects on a wide range of human embryonic development pathways. The mESC-EGFP assays with these pathway-specific gene promoters as molecular biomarkers thus can be used for HT screening of embryotoxic chemicals. The data obtained from the screening can provide the information needed for the regulation of drugs and chemicals.

630795 Developing a Photonics Platform for Determining Transport Rates of Proton Antiporters

Abstract:

Proton motive force is an important form of energy that drives many cellular processes including ATP production, membrane transport, as well as drug and antibiotic efflux. Therefore, many clinical targets are proton antiporters. The purpose of this work is to develop a photonics platform for determining molecular transport rates of proton antiporters. One such transporter complex, AcrAB can be found in *E. coli* and is a multidrug efflux pump that can contribute to antibiotic resistance. Using stopped flow fluorescence spectroscopy to track initial fluorescence change and fluorescence correlation spectroscopy to determine protein concentration and directionality, we have developed a methodology for quantitatively determining rates of both direct and indirect transport through the AcrAB complex. With the methodology developed, this platform can be used in the future to screen for inhibitors and be translated to other proton antiporters.

631071 Therapeutically Exploring Persister Metabolism in Bacteria

Abstract:

Bacterial persisters represent a small subpopulation of phenotypic variants in isogenic cell populations that are temporarily tolerant to high concentrations of antibiotics. The rise of antibiotic tolerance is one of the most critical global public threats of our century. Bacterial persistence contributes to this problem because these cells are thought to facilitate the re-occurrence of chronic infections (1, 2). Although persister eradication holds the potential to cure these infections, effective anti-persister methods remain scarce. We previously showed that persister cells were mostly obtained from stationary-phase-cell subpopulations with increased redox activities (3). These redox activities were largely maintained by degradation of endogenous proteins and RNA. This intracellular degradation yielded cell subpopulations that were less fit to produce protein and resume growth upon exposure to fresh nutrients, while providing temporary protection against antibiotics. Our study further demonstrated that inhibiting stationary-phase respiratory activities could prevent intracellular degradation and reduce persister formation.

In our current study, when we treated the stationary-phase cells with an ATP synthase inhibitor, chlorpromazine hydrochloride (CPZ) (4), we were able to reduce (more than 200-fold) both ampicillin and ofloxacin persistence in *Escherichia coli* (Fig. 1A). This pretreatment decreased stationary-phase-redox activities (Fig. 1B) and protein degradation, and gave rise to cells that succumbed to death upon exposure to antibiotics in fresh media. We note that CPZ is an FDA-approved, anti-psychotic drug that is effective, safe and listed as an essential medicine by the World Health Organization. To further identify additional anti-persister therapeutics that are medicinally relevant, we developed a rapid and straightforward chemical screening strategy using a degradable fluorescent protein and a small chemical library containing FDA-approved drugs and antibiotics among ~360 known chemical compounds. Our screening strategy identified several chemical inhibitors such as, polymyxin B, poly-l-lysine and phenothiazine anti-psychotic drugs, that were able to significantly reduce stationary-phase-redox activities and persistence in *E. coli* (Fig. 1A and B). Using fluorescent protein dilution method, these pretreatments also reduced viable but non-culturable cell (VBNC) formation during stationary phase. Due to their non-proliferating state, VBNC cells can also tolerate antibiotic treatments. However, unlike persisters, VBNC cells are rarely recolonize in standard culture medium in the absence of antibiotics. Pretreatment of stationary-phase *Pseudomonas aeruginosa* cultures with these chemicals also substantially reduced persistence, confirming the presence of a persistence mechanism similar to that found in *E. coli* (Fig. C). *P. aeruginosa* is involved in many hospital-related biofilm infections; it is the predominant cause of morbidity and mortality in cystic fibrosis patients with compromised immune systems (5).

Overall, our results demonstrate that our method can identify medically relevant chemicals. Our study further verifies that persister-cell metabolism is a rich source of therapeutic strategies to eliminate antibiotic-tolerant cells. This study has been published in "Frontiers in Microbiology" (6).

References

1. Lewis, K. (2007). Persister cells, dormancy and infectious disease. *Nature Reviews Microbiology*, *5*(1), 48-56.
2. Lewis, K. (2010). Persister cells. *Annual review of microbiology*, *64*, 357-372.
3. Orman, M. A., & Brynildsen, M. P. (2015). Inhibition of stationary phase respiration impairs persister formation in E. coli. *Nature communications*, *6*, 7983.
4. Bullough, D. A., Kwan, M., Laikind, P. K., Yoshida, M., & Allison, W. S. (1985). The varied responses of different F1-ATPases to chlorpromazine. *Archives of biochemistry and biophysics*, *236*(2), 567-575.
5. Mulcahy, L. R., Burns, J. L., Lory, S., & Lewis, K. (2010). Emergence of Pseudomonas aeruginosa strains producing high levels of persister cells in patients with cystic fibrosis. Journal of

bacteriology, 192(23), 6191-6199.
6. Mohiuddin, S. G., Hoang, T., Saba, A., Karki, P., & Orman, M. (2020). Identifying Metabolic Inhibitors to Reduce Bacterial Persistence. *Frontiers in Microbiology*, *11*, 472.

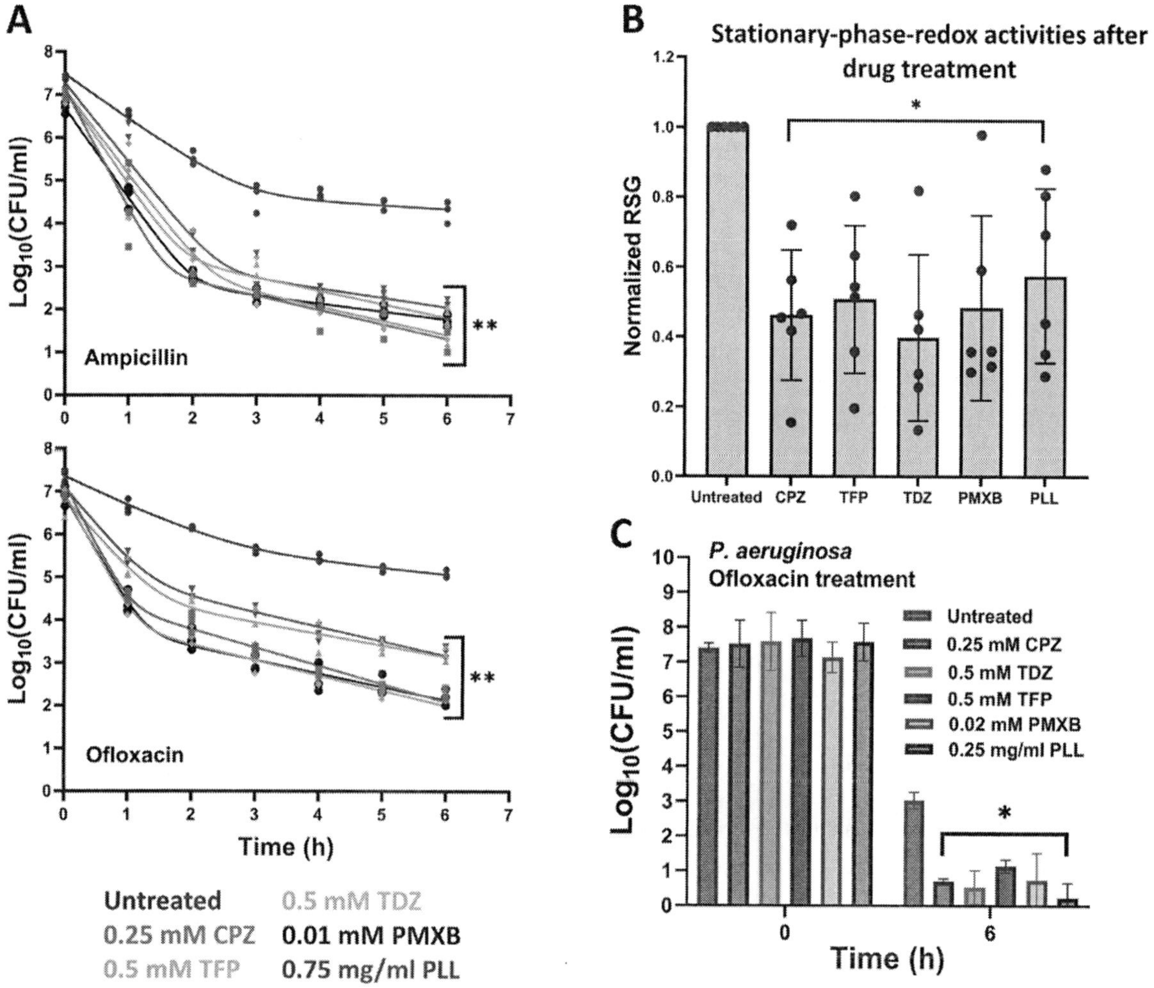

Fig. 1: Chemical treatments reduced stationary-phase persistence and redox activities in *E. coli*, and persistence in *P. aeruginosa* (A) Cells at early stationary phase (t=5 h) were treated with hit chemicals to inhibit the stationary phase metabolism. At late stationary phase (t=24 h), cells were washed to remove chemicals and diluted (1:100-fold) in fresh media and treated with ampicillin or ofloxacin to enumerate persisters (N=3). **(B)** Pretreated *E. coli* cells were stained with Redox Sensor Green (RSG) dye at t=24 h to measure their redox activities (N=6). **(C)** *P. aeruginosa* cultures were similarly treated with identified chemicals to test their impact on *P. aeruginosa* persistence. Cells were plated for colony forming unit (CFU) enumeration before and 6 hour after the ofloxacin treatments to assess the effects of drugs on *P. aeruginosa* cell viability and persistence, respectively
*Statistical significance between drug-treated and untreated cultures ($P < 0.05$, one-way ANOVA with Dunnett's posttest); ** Statistical significance between control (untreated) and treatment groups (P-value<0.0001, F-Statistics). CPZ, Chlorpromazine hydrochloride; PMXB, Polymyxin B; PLL, Poly-L-lysine; TDZ, Thioridazine; TFP, Trifluoperazine. N= Number of biological replicates.

627822 Development of a Novel Continuous Spatially Distributed Diafiltration Unit Operation

Abstract:

Flow chemical and continuous processes have seen significant uptake in R&D and small intensified manufacturing for complex chemical and pharmaceutical products. However, continuous work-up and purification operations for continuous manufacturing requires further development to approach the significant level of purification and stream adjustment in multistep routes that batch synthesis traditionally offers in the production of these classes of chemical products.

The objective of this study is to develop an operation that can conduct separations based on diafiltration using semipermeable nanofiltration or ultrafiltration membranes in a fully continuous manner in a single stage configuration. To this end a Continuous Spatially Distributed Diafiltration (CS2D) operation was developed, to conduct straight through diafiltration in a manner that would yield equivalent or better purification efficiency, yield and productivity to batch diafiltration. To achieve this goal diafiltration solvent is introduced spatially in a highly uniform manner with the flow guided by a range of 3D-printed mixers designed using CFD informed design to increase localized mixing of retentate and diavolumes flows or displacement effects within the channel. Static mixers were 3D-printed using titanium and polyether ether ketone (PEEK) providing high pressure and chemical compatibility (Figure 1), suitable for many flow chemical separations.

Figure 1: (a) mixer filled membrane flatform (1-bottom plate; 2- middle plate to hold membrane with double O-ring; 3-mixer; 4-top plate) and (b) selection of printed mixers

Ibuprofen was selected as a model active pharmaceutical ingredient (API) with methanol acting as a solvent and ethanol used as a model impurity where an organic solvent nanofiltration membrane was used to selectively retain ibuprofen. Experiments were processed at a 55 bar transmembrane pressure at which 87% and 15% rejection coefficients of ibuprofen and ethanol were determined respectively, in screening experiments. Using 4 diavolumes in constant volume diafiltration in batch and CS2D runs, purities of 55% and 76% were obtained at equivalent yields of approximately 61% for the Batch and CS2D separation respectively. Demonstrating that continuous spatially distributed diafiltration (CS2D) can match or exceed results conducted in equivalent batch configurations. As such CS2D may present an attractive intermediate purification and solvent swap operation for telescoped flow chemical and continuous processing applications, in addition to a highly compatible platform for use in automated flow-based experimentation.

Acknowledgements

This publication is supported SSPC, the Science Foundation Ireland Research Centre for Pharmaceuticals (12/RC/2275_P2) and I-form the Science Foundation Ireland Centre For Advanced manufacturing (16/RC/3872) and TIDA, SFI/EI Technology Innovation Development award (16/TIDA/4003) co-funded under the European Regional Development Fund.

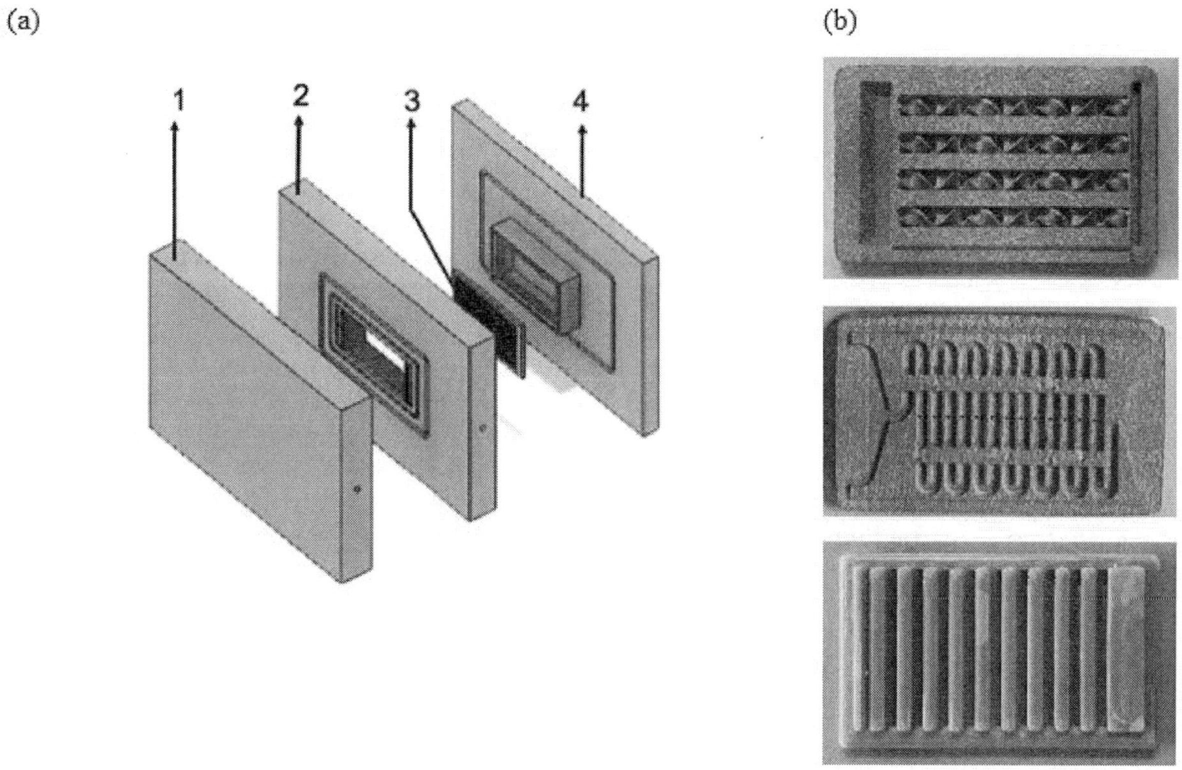
(a)
(b)
1
2
3
4

632591 Flow Chemistry As an Enabling, Green Technology in Merck Process

Abstract:

Herein we describe the use of flow chemistry in the development of a green reaction. Previous versions of the synthetic route had many chemical steps resulting in low yield and process mass intensity. A new flow process was developed that telescoped several steps into one reaction while utilizing cheap, green precursors. Performing the reaction in continuous flow minimized process risk and transfer unit operations while eliminating the need for in-process monitoring and engineering controls to suppress impurity formation. Data rich experimentation led to quick optimization of the flow reaction conditions within one month. This flow process was then successfully scaled and demonstrated at kilo and pilot plant scale.

629655 Impurity Purging through Systematic Process Development of a Continuous Two-Stage Crystallization

Abstract:

Continuous downstream processing of pharmaceutical products is often conducted in parallel with upstream development and optimization, leading to scarce and ever-changing input material. In order to efficiently make progress, meet deliverable deadlines, and match FDA requirements for API analysis, a systematic development plan must be imposed. Ciprofloxacin is included in the essential lists of medicines, and is the drug used to validate the POD (Pharmacy on Demand) system discussed in this report. Purification in this system is performed in a two-stage continuous MSMPR set-up with a first-stage pH-mediated precipitation, followed by a traditional seeded, anti-solvent crystallization second stage. Development of both stages was vastly conducted in batch to conserve material, while flow experiments were used for validation. OFAT and DOE approaches were used throughout to optimize the impurity purging. During process development a large change in input quality was achieved by switching a reagent in the final step of synthesis, leading to a different and improved impurity profile. To re-assess the purging ability of our system, more OFATs were conducted across temperature, solvent ratios, pH, and flow rates, to help re-optimize with the changed input across both stages. Upon optimization, it was confirmed that FDA specifications (0.07 % area by LC or less per impurity, 98-102% by mass assay) could be met. In finding this, one individual impurity remained at all close to the threshold and a majority of its purging was achieved in the first stage. To assess purging power of this stage, pH and temperature were varied as OFATs and in a 2 factor DOE and solid and liquid composition of API and this impurity were assessed. This impurity was found to have a linear relationship between purging and temperature that could be modeled. This linear model was then tested in flow to show errors of less then 40% consistently in predicting the purging levels of this individual impurity. By general process optimization this system was narrowed to necessitate control of a single impurity, which was found to be controlled quantitatively by temperature, ensuring product quality every time.

633279 Non-invasive, continuous, quantitative detection of powder level, mass holdup and moisture fraction in pharmaceutical GMP vessels

Abstract:

We describe a quantitative, non-invasive, GMP-compatible capacitance technique that enables implementation of continuous processing control in pharmaceutical metallic vessels. To illustrate the technique, we describe a probe producing a rapid signal that varies continuously with powder level and/or mass holdup in the metal feed tube of a tablet press. We outline the principle of the processing electronics, provide an algorithm for the accurate interpretation of its signals, pose general design guidelines for embedding the instrument in metal vessels, and show how the technique can be paired with another non-invasive reference probe to operate with varying powder blend and moisture. In this example, we illustrate a control strategy that keeps powder level constant in the feed tube. We also show how the same technique can be used to monitor concentration of water or other solvents by exploiting the corresponding changes in dielectric properties of pharmaceutical blends.

632421 Residence Time Distribution (RTD) Study of Chute Transition Zones between Unit Operations

Abstract:

Since continuous pharmaceutical manufacturing lines are unique to the manufacturing site, each manufacturing line can be wildly different even within the same company. One crucial tool utilized to ensure that the critical quality attributes of the final product in a continuous manufacturing (CM) line is the knowledge of the traceability or the residence time of material within the manufacturing line. Residence time distribution (RTD) is the measurement of time that a given amount of material resides in either a specific unit operation or throughout the entire continuous manufacturing line. Most of the RTD research concentrated on the RTD of the unit operations of CM and have either lumped in the connective parts, so they are included in a specific CM unit operation or have ignored the connection zones between each unit [1-6].

Since the main difference between the many different manufacturing lines available is transition zones, the work presented here has focused on traceability of material in transition zones to understand the effect of the distances between unit operations on the residence time distribution. In an integrated continuous manufacturing line, a blend of Compap L, Prosolv HD90, AcDiSol, and MgSt was dispensed into a transition chute where the hold up in the chute was controlled at three different levels. A pulse of Metformin was used as a tracer for the system. At the base of the set-up, in-line Near-Infrared detection was utilized to detect the tracer as it transitioned through the chute. The effect of critical process parameters was also investigated, such as throughput and blender speed, to capture a more detailed understanding of how the RTD is affected by the critical process parameters, specifically the transition zones.

Next, the effect of the tracer drop zone was investigated. The height at which the tracer falls into the system will affect the broadness of the RTD curve. Therefore, the height at which the tracer was introduced to the process stream was independently investigated. Since the material properties of the tracer can affect how the material falls, specifically as it displaces the air in the chute, three materials with a wide range of properties (i.e., cohesion, particle size, flowability) were selected as tracers to be introduced at different drop heights. Mass throughput was investigated to determine how the speed of the powder may exacerbate the dispersion. In all cases, a predicted NIR model was utilized to predict the concentration of the tracer as it transitioned passed the NIR probe. Using the concentration profile generated by the predictive model, the RTD parameters were regressed from the experimental data and compared using ANOVA to determine the effect of the critical process parameters. The results show that transition zones play a large part in RTD in a continuous line, greatly delaying pulse detection. In addition, the drop height of the pulse into one such transition zone can introduce noise into the residence time distribution, for which should be accounted.

References

1. Escotet-Espinoza, M. Sebastian, et al. "Effect of tracer material properties on the residence time distribution (RTD) of continuous powder blending operations. Part I of II: Experimental evaluation." *Powder Technology*342 (2019): 744-763.
2. Escotet-Espinoza, M. Sebastian, et al. "Effect of material properties on the residence time distribution (RTD) characterization of powder blending unit operations. Part II of II: Application of models." *Powder Technology*344 (2019): 525-544.

3. Engisch, William, and Fernando Muzzio. "Using residence time distributions (RTDs) to address the traceability of raw materials in continuous pharmaceutical manufacturing." *Journal of pharmaceutical innovation*1 (2016): 64-81.
4. Karttunen, Anssi-Pekka, et al. "Measurement of residence time distributions and material tracking on three continuous manufacturing lines." *International journal of pharmaceutics*563 (2019): 184-197.
5. Furukawa, Ryoichi, Ravendra Singh, and Marianthi Ierapetritou. "Effect of material properties on the residence time distribution (RTD) of a tablet press feed frame." *International Journal of Pharmaceutics*591 (2020): 119961.
6. Dülle, M., H. Özcoban, and C. S. Leopold. "The effect of different feed frame components on the powder behavior and the residence time distribution with regard to the continuous manufacturing of tablets." *International journal of pharmaceutics*555 (2019): 220-227.

626480 Adaptive Nonlinear Model Predictive Control of a Continuous Direct Compaction Tablet Manufacturing Process

Abstract:

Active control strategies are playing a vital role in modern pharmaceutical manufacturing. Indeed, digitalization and automation are revolutionizing the pharmaceutical industry and are accelerating the shift from batch operations to continuous operations. To deal with variations in raw material properties and process uncertainties as well as to guarantee desired critical quality attributes (CQAs) of in-process materials and final products, active control strategies are needed to provide real-time corrective actions when departures from quality targets are detected. Instead of applying traditional control strategies by controlling process input variables at fixed setpoints or within tight ranges, active control strategies serve as a "must-have" technology, as mentioned in the recent AIChE PD2M Advanced Process Control workshop.[1]

A three-level hierarchical control structure has been applied in the pilot plant at Purdue University in order to achieve effective setpoint tracking and disturbance rejection in the tablet manufacturing process.[2] Level 0 control is equipment built-in control, Level 1 control is process analytical technology (PAT)-based PID control, and Level 2 control is optimization-based model predictive control (MPC). As can be expected, when the multiple-input multiple-output (MIMO) system exhibits strong process variables interaction, MPC demonstrates better control performance than classical PID control. That is because MPC can predict the transition of controlled variables (e.g. API content, tablet weight) instantaneously and optimize the control moves of all the manipulated variables (e.g. API flow rate, depth of filling) within a finite time horizon. Several MPC implementation case studies have been reported: (1) the API concentration can be controlled in the feeding-blending unit,[3] (2) the compaction force and the tablet weight are controlled in a rotary tablet press, [4-5] and (3) the hopper level of the tablet press and the API concentration are controlled in a direct compaction line.[6] While satisfactory control performance is found in the aforementioned research studies, the implementation of MPC in actual pharmaceutical manufacturing processes is still in the infant stage.

This work is focused on developing and implementing nonlinear model predictive control (NMPC) in continuous tablet manufacturing. To mitigate the control degradation caused by plant-model mismatch, careful monitoring and continuous improvement strategies are required. When moving horizon estimation (MHE) is integrated with NMPC, historical data in the past time window together with real-time data from the sensor network, enable state estimation and model parameter updating. The adaptive model in the NMPC strategy can compensate for the process uncertainties, further reducing plant-model mismatch effects. The nonlinear first-principles model used in both HME and NMPC can predict the essential but complex powder properties and provide physical interpretation of abnormal events. In the continuous direct compaction process studied in this work, CSTR-in-series models are used to describe the dynamics of the feeding-blending unit and the Kawakita model accounts for powder compressibility in the tablet press. The adaptive NMPC implementation and its real-time control performance analysis will be demonstrated on our continuous tableting pilot plant.

References

[1] Huang, J., O'Connor, T., Ahmed, K., Chatterjee, S., Garvin, C., Ghosh, K., ... & Warman, M. (2021). AIChE PD2M Advanced Process Control workshop-moving APC forward in the

pharmaceutical industry. Journal of Advanced Manufacturing and Processing, 3(1), e10071.

[2] Su, Q., Ganesh, S., Moreno, M., Bommireddy, Y., Gonzalez, M., Reklaitis, G. V., & Nagy, Z. K. (2019). A perspective on Quality-by-Control (QbC) in pharmaceutical continuous manufacturing. Computers & Chemical Engineering, 125, 216-231.

[3] Singh, R., Sahay, A., Karry, K. M., Muzzio, F., Ierapetritou, M., & Ramachandran, R. (2014). Implementation of an advanced hybrid MPC–PID control system using PAT tools into a direct compaction continuous pharmaceutical tablet manufacturing pilot plant. International journal of pharmaceutics, 473(1-2), 38-54.

[4] Su, Q., Bommireddy, Y., Shah, Y., Ganesh, S., Moreno, M., Liu, J., ... & Nagy, Z. K. (2019). Data reconciliation in the Quality-by-Design (QbD) implementation of pharmaceutical continuous tablet manufacturing. International journal of pharmaceutics, 563, 259-272.

[5] Bhaskar, A., Barros, F. N., & Singh, R. (2017). Development and implementation of an advanced model predictive control system into continuous pharmaceutical tablet compaction process. International journal of pharmaceutics, 534(1-2), 159-178.

[6] Kirchengast, M., Celikovic, S., Rehrl, J., Sacher, S., Kruisz, J., Khinast, J., & Horn, M. (2019). Ensuring tablet quality via model-based control of a continuous direct compaction process. International journal of pharmaceutics, 567, 118457.

629216 Modeling Impurity-Mediated Crystal Growth

Abstract:

Crystallization is the process by which growth units, such as molecules and ions, organize to form a structured solid state. High purity crystals grow via a layer-by-layer process known as spiral growth, which is fundamental to various applications including the development and manufacture of pharmaceuticals, catalysts, and organic light-emitting diodes (OLEDs). Given the ubiquity of crystal growth in industrial processes, there is substantial demand for predictive and mechanistic modelling of crystallization. Crystallization of organic molecules is well understood for ideal systems--i.e., Kossel crystals with a single centrosymmetric growth unit (simple cubic single molecules with equal surfaces). There is interest in studying crystal systems in which non-idealities are introduced, as these are more representative of realistic conditions. One such non-ideality involves the presence of impurities.

The goal of this work is to investigate the effect of impurities or 'imposter molecules' on the crystal growth process and to develop theoretical models for the mechanisms by which impurities influence crystal growth and hence affect crystal morphology and size. Herein, an impurity, or imposter, typically refers to a structurally similar species to, or one which inhibits the growth of, the primary crystallizing species. Structurally similar additives may influence crystallization of active pharmaceutical ingredients, such as paracetamol. While these systems are prevalent in many relevant fields, there is currently no comprehensive model that captures the full effect of impurities on crystal growth processes. A deeper understanding of the underlying growth mechanisms would promote control of crystallization processes via tailor-made additives and enable identification and proper treatment of impurities that alter growth rates. We employ Kinetic Monte Carlo (KMC) to simulate the time evolution of centrosymmetric organic crystal growth and developed model equations to quantify the effect of inhibition on the crystal face for systems populated with impurities.

In order to validate these simulations, we have developed a classification algorithm to distinguish amongst distinct crystal edge sites on the moving step. We have compared the densities of such sites from simulation results to predictions according to kink density expressions such as the Frenkel expression as well as more thorough site probability expressions developed by Lovette and Doherty (Phys. Rev. E **2012, 85**, 021604). We observe excellent agreement between the two approaches which we take as validation of both.

The validated KMC simulations were then used to test the classical mechanistic step velocity model which states that step velocity is proportional to the product of kink density and kink rate (net attachment rate of growth units into kink sites). The agreement is demonstrated to be very good over a range of bond energies and supersaturations, thus validating the classical mechanistic approach. Given the agreement over various growth conditions, we then incorporate impurities in the KMC simulation methodology to investigate their effect on crystal growth. For the cases studied, our results generally suggest a decrease of the step velocity with increasing impurity concentration, though the form of this dependence is based on the mobility and distribution of the impurities, as well as the energetic interactions between impurities and the primary crystallizing species. We use these model equations to predict morphology changes for adipic acid grown from aqueous solution. These results are visualized within the framework of ADDICT, a practical engineering tool developed in the Doherty group to compute relative growth rates and visualize morphology of solution-grown faceted crystals on the order of seconds.

624763 Combination of in Situ Imaging and Deep Learning for Crystallization Process Monitoring: Application to Beta-Lactam Antibiotics Manufacturing

Abstract:

Many crystallization processes at some point encounter undesired solid species, whether they be polymorphs, byproducts, or crystals with undesired shapes. Detection and quantification are challenging tasks when undesired particles make up a small fraction (e.g., < 2%) of the suspended solids. Fault detection would save resources if a batch could be rejected following the appearance of an undesired solid rather than at the end of the process, or feedback control strategies can be implemented to eliminate the undesired particles.

In situ imaging is commonly used to monitor different crystallization processes, characterize the size and shape of product, and detect the presence of undesired shapes or polymorphs [1,2]. There are a series of challenges that hinder the automation of analyzing the captured frames, which is necessary for long-term operation, such as in continuous processes. One challenge is imaging itself; optically dense slurries do not lend themselves to in situ microscopy. Often, *ex situ* sample loops are employed, which utilize hydrodynamic focusing, backlighting, and dilution to obtain images of in-focus, clearly delineated, non-overlapping crystals [2]. In these cases image quality may be gained at the cost of not fully representing the population and conditions in the sample loop may not match those in the crystallizer. Besides practical problems associated with having a camera inside or in a loop parallel to a vessel, a range of challenges related to the analysis of captured images may prevent gaining usable information from the images. Some of the common challenges in image analysis are illumination, rotation, deformation, overlap, and background clutter.

In this work, we develop a computer vision model based on convolutional neural networks (CNN) that can be combined with high-throughput *in situ* imaging for monitoring a crystallization process. We are interested in characterizing the state of a system with optically dense slurries where *in situ* images contain highly over-lapped objects and are hard to analyze with traditional approaches. Our goal is to develop a tool for automated process monitoring that can (1) use *in situ* images to monitor the state of the process, (2) detect the presence of undesired particles in the slurry, and (3) estimate the location of the undesired particle for further inspection and taking proper control actions. We focus on the application of such a tool in continuous pharmaceuticals manufacturing, where robust monitoring is essential for the production of a pure and high-quality product. We take the production of cephalexin monohydrate (a beta-lactam antibiotic with a large annual production rate of more than 5,000 tons per year [3]) as an example where a byproduct (phenylglcyine) of the reaction may crystallize and contaminate the product in the event of process deviations from the setpoint [4].

Using more than 8000 images, we re-trained the *ResNet-18* CNN for the classification of two classes, cephalexin and phenylglcyine particles (the former are needle-like but latter more spherical blobs) with more than 98% accuracy on the validation data [5]. This network is then used to inspect different regions of each image captured from the crystallization vessel using a sliding window approach to detect the presence of phenylglcyine particles. Our analysis of dozens of images shows that the developed model can detect the presence and location of the undesired particles with high accuracy (Fig. 1). Besides discussing our results for this case and a second CNN (*ResNet-50*), we compare the performance of both models in a range of solid concentrations. We also describe a straightforward procedure for extending this methodology to other crystallization systems using open-source deep learning development frameworks such as TensorFlow.

Fig. 1 The overall process of detecting a PG crystal: A sliding window scans the image, cropping the proposed region in the raw image, passing it through the CNN, generation of feature maps, and classification (top). Results of applying the image analysis model (with ResNet-50 CNN) to a series of in situ images acquired from a crystal slurry (bottom). Solid red boxes correspond to regions that are decided to have a PG particle with 0.8 classification cutoff.

Cited literature

[1] Simon, Levente L., Hajnalka Pataki, György Marosi, Fabian Meemken, Konrad Hungerbühler, Alfons Baiker, Srinivas Tummala et al. "Assessment of recent process analytical technology (PAT) trends: a multiauthor review." Organic Process Research & Development 19, no. 1 (2015): 3-62.

[2] Liu, Tao, Yan Huo, Cai Y. Ma, and Xue Z. Wang. "Sparsity-based image monitoring of crystal size distribution during crystallization." Journal of Crystal Growth 469 (2017): 160-167.

[3] Wegman, Margreth A., Michiel HA Janssen, Fred van Rantwijk, and Roger A. Sheldon. "Towards biocatalytic synthesis of β-lactam antibiotics." Advanced Synthesis & Catalysis 343, no. 6-7 (2001): 559-576.

[4] Salami, Hossein, Colton E. Lagerman, Patrick R. Harris, Matthew A. McDonald, Andreas S. Bommarius, Ronald W. Rousseau, and Martha A. Grover. "Model development for enzymatic reactive crystallization of β-lactam antibiotics: a reaction–diffusion-crystallization approach." Reaction Chemistry & Engineering 5, no. 11 (2020): 2064-2080.

[5] Salami, Hossein, , Matthew A. McDonald, Andreas S. Bommarius, Ronald W. Rousseau, and Martha A. Grover. "In situ imaging combined with deep learning for crystallization process monitoring: Application to cephalexin production." Organic Process Research and Development (submitted)

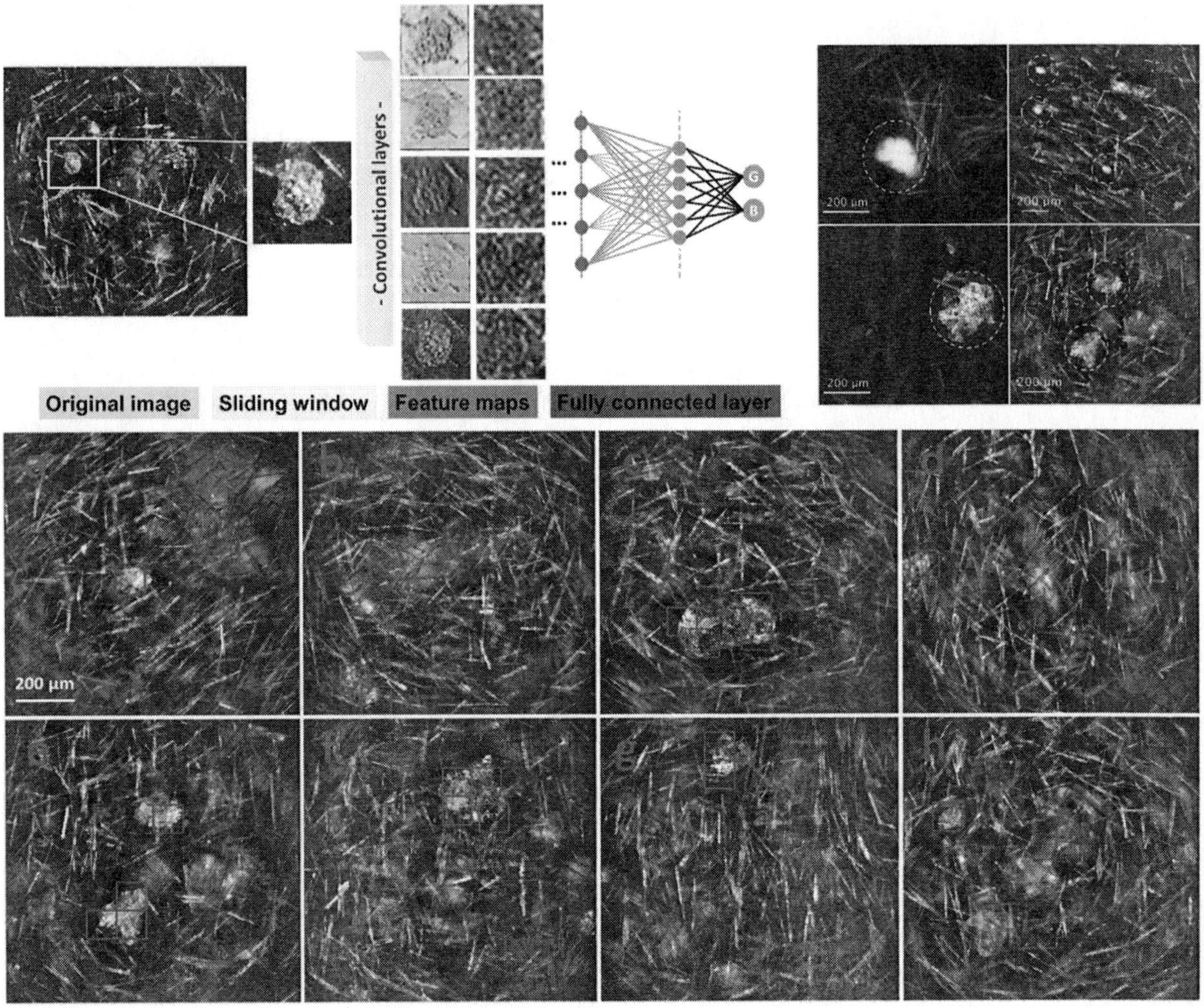
- Convolutional layers -
G
B
Original image
Sliding window
Feature maps
Fully connected layer
200 µm
200 µm
200 µm
200 µm
200 µm

630706 Brownian Dynamics Simulations Predict the Nucleation Probability of Polymorphs during the Process of Crystallization

Abstract:

Crystallization of organic molecules is a highly stochastic process involving a large number of molecules spanning long timescales. Few degrees of freedom arising due to the molecular structure can result in multiple possible metastable crystal structures, leading to polymorphism. Such metastable structures prevent the ability to predict and control the outcome of crystallization effectively. To connect the events that occur during the molecular motion to the outcome of crystallization, the calculation of the probability of events leading to desired polymorphs is necessary. In previous studies, we have identified the rapid solvent exchange in the solvation shell and configuration of the solvent molecules in the solvation shell as the critical aspects of crystallization that lead to the desired polymorph. To further obtain the probability of events leading to the desired polymorph, we use the Brownian Dynamics simulation approach. Here, we coarse-grain the molecular structure to reduce the number of degrees of freedom, and the effect of solvent interaction is captured by the noise term in the Langevin equation. This approach allows simulating the interaction of a large number of solute molecules and spans a longer timescale. The configuration of solute molecules at each time is used to predict the probability of events leading to a particular polymorph. The approach stated above is applied to the Glutamic acid molecule to effectively predict the polymorphic form obtained at the end of the crystallization process. Furthermore, results suggest that an increase in the local density due to high supersaturation allows the formation of metastable polymorph at a greater probability, and tuning the solvent environment to restrict such events can yield significant control on the process of crystallization.

632740 Kinetic Monte Carlo Modeling of MIL-53 Metal Organic Framework Crystal Growth

Abstract:

Metal organic frameworks (MOFs) are crystalline, coordination polymers well-known for their applications in separations, catalysis due to their large surface areas and high porosity. These materials are structurally versatile because they are composed of metal cluster and organic ligand pairs, which can be strategically switched to make MOFs with certain functionalities. A subset of MOFs with a "rod-packing" motif ("rod-MOFs) are promising as breakthrough materials for size-selective, passive chemical separation processes due to their fine unidirectional channels. The main challenge for these applications is the optimization of the crystallization process for these MOFs, that is, to make them larger than their conventional micron scales and with negligible intergrowths.

Crystallization through the hydrothermal method is a common process used to synthesize certain classes of metal organic frameworks (MOFs), including MIL-53, which falls under the rod-MOF classification. Though the process is ubiquitous to the formation of a variety of crystals, control of the dynamics of the growth process itself remains a challenge as the closed nature of the batch system typically used for these reactions are reminiscent of a "black box". A wealth of literature exists on the exploration of crystal growth dynamics through Monte Carlo (MC) and Molecular Dynamics (MD) simulations, as well as powerful models which incorporate both. These models have seldom been applied to investigate the crystal growth kinetics of MOFs, and more specifically, the MOF crystallization phenomena for special cases where the inputs to the system are changing periodically.

An understanding of MOF crystallization phenomena under hydrothermal conditions from a fundamental perspective is emphasized herein through the application of kinetic Monte Carlo algorithms (kMC) based on the solid-on-solid (SOS) model.

627019 Digital Twin Design of an Agrochemical Crystallization Process Using a 2D Population Balance Model Approach

Abstract:

In the agrochemical industry, crystallization is a key separation process that is used in the to separate the agrochemical actives from reaction mixtures and impure solutions. A poorly designed crystallization step can lead to poor yield, low purity, and long filtration time. To minimize process cycle time, a series of design of experiments led by the Quality-by-Design (QbD) approach can be used by practice to optimize the operating profile of a crystallization process (i.e. temperature profile, solvent/antisolvent addition profile, pH profile).[1] However, an exhaustive list of experiments covering all factors is needed to explore the whole design space experimentally. To avoid excessive open-loop experiments and personnel exposure to toxic chemicals, recently introduced Quality-by-Control (QbC) methodology controls critical quality attributes (CQAs) through implementation of feedback control algorithm to determine the operating profile of the process.[2] For rapid process design, direct design or model free (mfQbC) approaches can be used to generate a robust operating profile leading the system to the desired CQAs. However, mfQbC results are often suboptimal with new process changes (scale-up, mixing conditions, and solvent addition), which requires model-based QbC (mbQbC) to further optimize the mfQbC results.[3,4]

In this work, a digital twin of an industrial agrochemical crystallization process was developed to generate a temperature cycling operating profile with the objective of maximizing crystal size to reduce filtration time. A 2D population balance model (PBM) solved via semi-discrete finite volume method was developed. Parameter estimation experiments to determine the 2D kinetics was performed using a well-mixed 2D PBM and laboratory scale experiments. Scale-up and antisolvent addition effects were incorporated in the digital twin by using a multicompartmental PBM approach with parameters refined from large-scale experiments. Various compartments were required to capture wall-effects from large jacket to vessel temperature differences and high shear environment near the impeller area.

References:

(1) Bondi, R. W.; Drennen, J. K. *Quality by Design and the Importance of PAT in QbD*; Academic Press, 2011; Vol. 10. https://doi.org/10.1016/B978-0-12-375680-0.00005-X.

(2) Su, Q.; Ganesh, S.; Moreno, M.; Bommireddy, Y.; Gonzalez, M.; Reklaitis, G. V.; Nagy, Z. K. A Perspective on Quality-by-Control (QbC) in Pharmaceutical Continuous Manufacturing. *Comput. Chem. Eng.* **2019**, *125*, 216–231. https://doi.org/10.1016/j.compchemeng.2019.03.001.

(3) Szilagyi, B.; Eren, A.; Quon, J. L.; Papageorgiou, C. D.; Nagy, Z. K. Application of Model-Free and Model-Based Quality-by-Control (QbC) for the Efficient Design of Pharmaceutical Crystallization Processes. *Cryst. Growth Des.* **2020**, *20* (6), 3979–3996. https://doi.org/10.1021/acs.cgd.0c00295.

(4) Öner, M.; Bach, C.; Tajsoleiman, T.; Molla, G. S.; Freitag, M. F.; Stocks, S. M.; Abildskov, J.; Krühne, U.; Sin, G. Scale-up Modeling of a Pharmaceutical Crystallization Process via Compartmentalization Approach. In *Computer Aided Chemical Engineering*; Elsevier B.V., 2018; Vol. 44, pp 181–186. https://doi.org/10.1016/B978-0-444-64241-7.50025-2.

625819 Optimization of Energetics Crystallization Via Quality-By-Control Direct Design Approaches

Abstract:

Crystallization is an essential process of solids manufacturing and is left unoptimized in several fields, including energetics manufacturing. Unoptimized crystallization protocols can lead to solid material with undesired physical and chemical characteristics, such as particle morphology, stability, detonation potential, manufacturability, and overall crystal quality. Although researchers have extensively studied the solubility of the common energetic materials, Research Department/Royal Demolition Explosive (RDX) and High Melting Explosive (HMX), little work has been completed on the optimization of both batch and continuous crystallization processes for RDX and HMX.

In this work we demonstrate the application of the Quality-by-Control (QbC) framework for the rapid design of the crystallization of energetic compounds. QbC allows for the minimization of experiments and experimental exposure by utilizing feedback control strategies for the critical quality attributes (CQAs).[1] Using the QbC-based direct design approach an optimized method for the industrial manufacturing of crystallized RDX and HMX could be suggested. Our direct design approaches require initial solubility and solvent selection data. Small-scale experiments completed with Crystal16 showed that RDX and HMX have high solubility in γ-Butyrolactone and a sizeable metastable zone width (MSZW), which is desirable for crystallization control. The two applied direct design approaches are direct nucleation control (DNC) and supersaturation control (SSC). DNC applies particle counts measurements collected with an *in-situ* focused beam reflectance measurement (FBRM) probe to a model-free closed feedback control loop which introduces temperature cycling to a saturated solution, stimulating controlled nucleation and crystal growth.[2] SSC applies solution concentration measurements collected with an *in-situ* infrared (IR) probe to a model-free closed feedback control loop which introduces temperature cycling to a saturated solution, stimulating controlled solution concentrations.[3] In application with *in-situ* process analytical technology (PAT) tools, DNC and SSC allowed for the optimization and selection of crystallization techniques for RDX and HMX at the research-scale. Understanding an optimal method for the industrial crystallization of common energetic materials, such as RDX and HMX, will help improve the quality and effectiveness of manufactured energetic materials and enable future process development for the manufacturing of energetic materials.

References

1. Su, Q.; Ganesh, S.; Moreno, M.; Bommireddy, Y.; Gonzalez, M.; Reklaitis, G. V.; Nagy, Z. K. A Perspective on Quality-by-Control (QbC) in Pharmaceutical Continuous Manufacturing. *Chem. Eng.* **2019**, *125*, 216–231.

2. Bakar, M. R. A.; Nagy, Z. K.; Saleemi, A. N.; Rielly, C. D. The Impact of Direct Nucleation Control on Crystal Size Distribution in Pharmaceutical Crystallization Processes. *Growth Des.* **2009**, *9* (3), 1378–1384.

3. Saleemi, A. N.; Rielly, C. D.; Nagy, Z. K. Comparative Investigation of Supersaturation and Automated Direct Nucleation Control of Crystal Size Distributions Using ATR-UV/Vis Spectroscopy and FBRM. *Growth Des.* **2012**, *12* (4), 1792–1807.

625783 Use of Bayesian Modeling for Failure Risk Analysis and Control Strategy Design

Abstract:

Process characterization and definition of appropriate control strategies is critical to ensure quality in the manufacture of active pharmaceutical ingredients. According to the principles of quality by design (QbD), process development should include identification of critical quality attributes (CQAs), characterization to understand the relationship between process parameters and these CQAs, and an evaluation of risk for the process. This presentation provides a case study wherein Bayesian probabilistic modeling was used to evaluate process risk for a specific critical impurity. Formation and purging of this impurity was studied via DoE and mechanistic studies spanning the proven acceptable ranges (PAR) for reaction and crystallization parameters. Probabilistic models were developed to predict process performance and assess and visualize predicted failure rates across the multivariate parameter space. This analysis informed assignment of critical process parameters and appropriate multivariate PARs to ensure robust processing. The probabilistic models also enabled estimation of the impact of common cause plant variability on the anticipated failure rate against specification levels of the critical impurity. Overall, these various failure rate analyses, enabled by the Bayesian models, helped to ensure implementation of an effective control strategy and will aid in future process transfers and process monitoring upon commercialization.

626120 Accelerated Process Design and Optimization for a Small Molecule COVID-19 Therapeutic

Abstract:

Molnupiravir is an investigational oral antiviral treatment for SARS-COV-2 that works as a ribonucleoside analog to inhibit replication of RNA viruses. The proposed commercial manufacturing route is a five step synthetic route from the regulatory starting material. This presentation will discuss the optimization of the final transformation on a highly accelerated timeline to ensure metric-ton supply of high purity molnupiravir during the coronavirus pandemic. This step begins with an acid-catalyzed deprotection reaction to form molnupiravir from the penultimate intermediate which also forms impurities due to acid-sensitive hydrolysis. Small-scale high throughput experimentation was used to build kinetic models which were used to optimize impurity generation and yield productivity. The workup of the reaction also introduces the risk of producing lightly purple colored active pharmaceutical ingredient (API) through trace metal coordination, which was mitigated through the addition of a chelating agent. The chelated metal and other aqueous-soluble impurities are rejected through extractions. Remaining impurities are then rejected into the mother liquor during the azeotropic distillative crystallization. Because of the potential variability of the distillation across several equipment trains and scales, the crystallization was designed for better control and less sensitivity to these variations. The resulting process performance has been monitored at four clinical and commercial manufacturing sites. In summary, this talk will discuss the development of control strategies across the last step of the molnupiravir synthesis to ensure the critical quality attribute of purity is met.

628959 A Digital Twin of Flexible Modular Continuous API Manufacturing Process

Abstract:

Advanced continuous manufacturing (CM) is emerging as a preferred platform for the production of active pharmaceutical ingredients (APIs). The future continuous pharmaceutical manufacturing is desired to be based on modular plug and play concept. In this approach, mobile modules consisting of a group of unit operations have been developed that can be easily reconfigured to assemble a scalable pharmaceutical manufacturing process for different products. It is also suitable for on-demand manufacturing such as in the case of COVID-19. However, because of different levels of complexity, the development and adaptation of flexible and modular CM plant is still an open area of research. Therefore, a systematic framework of methods and tools are needed to develop, adapt and evaluate the flexible modular CM plant for API manufacturing with reduced time and resources. In particular, a digital twin is critical for quick design, adaptation, optimization, scale up, and control of continuous APIs manufacturing process, allowing for virtual modeling of operations without extensive process development.

In this work, a digital twin for continuous APIs manufacturing process using modular components has been demonstrated. The process flowsheet model is the heart of 'digital twin' and consists of mathematical representation of three modules with the options of adding more as needed. The first module is for feeding and dispensing. It consists of a refill unit, feed tank, pump, mixing, and preheater. The second module is for performing the chemical reactions required to produce the target APIs. This module consists of tubular reactors placed inside a heating and cooling jacket. The third module is focused on separation of API from impurities. A continuous chromatographic model is currently used for separation purposes that can be easily replaced with any other type of separation techniques such as continuous crystallization. The applications of the digital twin have been demonstrated for design and development of suitable control architecture for a continuous APIs manufacturing process. The developed control architecture is also modular in nature and can be easily adapted for different manufacturing processes of APIs that may have different material and information flows, dead time variations, and tuning parameters. The proposed digital twin can save the time and resources needed for continuous APIs manufacturing and improve the product quality significantly.

The objective of this presentation is two-fold; first to highlight the developed digital twin and then demonstrate its application for design of process and control architecture of continuous API manufacturing processes.

624677 Quality-By-Control of a Novel Unit for Continuous Integrated Filtration-Drying of Drug Substances

Abstract:

Quality-by-Control (QbC) is a recent evolution [1] of the Quality-by-Design [2] initiative, in which active process control represents the core feature of the control strategy. With an active process control system in place, the critical process parameters of the process are automatically manipulated in response to disturbances, so that the target critical quality attributes of the product are ensured. The QbC framework has acquired even more importance within the recent interest in continuous integrated pharmaceutical manufacturing [3]. Continuous manufacturing (CM) guarantees greater product consistency compared to batch processing, as continuous processes present inherently greater process controllability. QbC has emerged as a very useful tool for the transition to CM, which requires a systematic framework for process control.

In this study, we develop a QbC framework for an intensified unit for continuous integrated filtration-drying of crystallization slurries. Continuous and integrated solutions for filtration and drying in upstream manufacturing are not much studied in the literature, although these processing steps represent a main obstacle for the implementation of end-to-end CM. The unit studied in this work is a novel carousel technology [4]. The carousel presents multiple processing stations located within a main cylindrical body. In the first station, the crystallization slurry is loaded. In the subsequent stations, slurry filtration, cake washing, cake deliquoring and thermal drying are carried out. The dry cake is discharged from the last station through the action of a pneumatic piston. All stations operate simultaneously, processing different batches of material: every fixed cycle duration the carousel rotates, enabling pseudo-continuous operation. Filtration, washing and drying operations present high coupling *per se*, and are even more interacting within the rotation mechanism of the carousel, which fixes the residence time of each processing step.

To aid the development of the control strategy for the unit, we develop a comprehensive digital twin of the carousel. The digital twin is based on detailed first-principles models of the processing steps [5], combined together for simulating real time operation of the carousel. The developed framework represents a unique and effective environment for testing and comparing different control strategies, since: *i)* time-consuming experiments with different control strategies can be quickly simulated on the digital twin, and *ii)* the critical quality attributes of the product, such as the solvents and impurities content, are always accessible in the *in-silico* simulations, while they cannot be measured in real time on the physical unit.

After having developed the digital twin, we validate it with filtration and drying experimental data collected on a physical carousel. Then, we conceive and implement on the digital twin a three-level control system for the carousel, based on a recently proposed approach [1] that follows the ISA-95 Enterprise-Control System Integration Standard. The Level 0 of the control system is made up of the built-in controllers of the carousel and of the accessory equipment, such as the pumping system. End-point control is resorted to at Level 1, for automatically setting the duration of each cycle to ensure that the discharged cake meets the target residual solvents and impurities content. Within the end-point control loop, the available measurements from the process are used to infer in real time the time instant at which the cake can be discharged because the desired quality has been met, and the carousel rotation routine is consequently automatically run. We also implement a model-based Level 2 control system, which includes a state estimator for continuous monitoring of the critical quality attributes and real time optimization. We successfully test the proposed control system in normal operating conditions and in presence of filter mesh fouling, which is known to be the main disturbance occurring during carousel operation.

References

[1] Su, Q., M. Moreno, C. Laird, Z. Nagy and G. Reklaitis (2019). A perspective on Quality-by-Control (QbC) in pharmaceutical continuous manufacturing, *Comput. Chem. Eng.*, **125**, 216–231.

[2] Food & Drug Administration, 2004. Guidance for industry, PAT-A Framework for Innovative Pharmaceutical Development, Manufacturing and Quality Assurance. http://www. fda. gov/cder/guidance/published. html. Accessed on 03/29/2021.

[3] Lee, S. L., T. F. O'Connor, X. Yang, C. N. Cruz, S. Chatterjee, R. D. Madurawe, C. M. V. Moore, L. X. Yu and J. Woodcock (2015). Modernizing Pharmaceutical Manufacturing: from Batch to Continuous Production. *J. Pharm. Innov.,* **10,** 191–199.

[4] Liu, Y.C., A. Domokos, S. Coleman, P. Firth, Z.K. Nagy (2019). Development of continuous filtration in a novel continuous filtration carousel integrated with continuous crystallization, *Org. Proc. Res. & Dev.*, **23** (12), 2655-2665.

[5] Destro, F., V. Wang, M. Abdi, X. Feng, E. Wood, E., S. Coleman, P. Firth, A. Barton, M. Barolo and Z. K. Nagy, (2021). Continuous integrated filtration, washing and drying of aspirin: digital design of a novel intensified unit. Proc. of the 11th IFAC Symposium on Advanced Control of Chemical Processes, Venice (Italy), June 13-16 2021. In press.

630559 Data Pre-Treatment Analysis of Residence Time Distribution (RTD) Profiles for Pharmaceutical Manufacturing Applications

Abstract:

Given the numerous advantages of continuous manufacturing (CM) [1] in the pharmaceutical industry, there has been an immense research thrust aimed towards understanding process dynamics [2] for accurate process prediction and efficient implementation of CM. Amongst the various techniques [2, 3] implemented for this end, the characterization of residence time distribution (RTD) has been performed for various unit operations as well as entire manufacturing lines, to capture the overall system dynamics followed by a change in the processing conditions [4]. RTD profile of any unit operation in the pharmaceutical manufacturing lines, can be obtained by adding a pulse or step change of the selected tracer at the inlet of the unit operation and measuring the tracer concentration at the outlet as a function of time using Process Analytical Technology (PAT).

However, the tracer concentration profile can include significant intrinsic and extrinsic noises from various sources. The different sources of "extrinsic noise" include environment and process conditions. The sources of "intrinsic noise", on the other hand, refer to the internal fluctuations of the system, which can be classified into the sensor, equipment, and material variability related to the inherent variability of the sensor during RTD measurement, equipment fluctuations during operation, and differences of tracer and bulk material properties, respectively [5, 6]. It is important to measure, quantify and remove the external noise as well as measurement variability from the RTD profiles obtained experimentally, as they can lead to inaccuracies in developing predictive models for evaluating system dynamics. However, the available literature focusing on the application of RTD for pharmaceutical manufacturing vaguely mentions enough details on this aspect of data pre-treatment of RTD profiles.

In this work, we aim to present and discuss various data pre-treatment and noise handling strategies aimed towards identifying effective strategies for denoising RTD profiles. The first aspect of the proposed work focuses on the quantification of the noise associated with the experimental RTD measurements. Following this, the second aspect focuses on handling the measurement variability and external noise for reliable RTD measurements. To this end, strategies similar to those available in image processing [7] and signal processing [8] are implemented. Some examples of noise filtering include linear and non-linear filters like adaptive filter, Kalman filter, recursive least-square, least mean square filter, and median filter. To reduce the variability, signal transforms, such as Fourier transform and Wavelet transform [9], are used for noise reduction, followed by low-pass or high-pass filtering. In the proposed work, the various data pre-treatment and noise handling strategies including time and frequency averaging, and exponential smoothing techniques, are evaluated and compared, to obtain a good signal-to-noise ratio (SNR) of the experimentally obtained RTD profile. Lastly, the effectiveness of different averaging and denoising techniques is analyzed to ensure that the important information pertaining to the RTD profile is retained. Thus, the proposed work aims towards developing guidelines for data pre-treatment strategies of RTD, which would be helpful towards effective utilization of RTD studies for predictive modeling of manufacturing lines.

References:

1. Schaber, S.D.e.a., *Economic Analysis of Integrated Continuous and Batch Pharmaceutical*

Manufacturing: A Case Study. . Ind. Eng. Chem. Res. , 2011. **50**: p. 10083–10092.
2. M. Ierapetritou, F.J.M., G. Reklaitis, *Perspectives on the continuous manufacturing of powder-based pharmaceutical processes.* AIChE Journal, 2016. **62**: p. 1846-1862.
3. V. Vanhoorne, C.V., *Recent progress in continuous manufacturing of oral solid dosage forms.* International Journal of Pharmaceutics, 2020. **579**: p. 119-194.
4. Danckwerts, P.V., *Continuous flow systems: Distribution of residence times.* Chemical Engineering Science, 1953. **2**: p. 1-13.
5. Kirsch, J.D. and J.K. Drennen, *Near-infrared spectroscopy: applications in the analysis of tablets and solid pharmaceutical dosage forms.* Applied Spectroscopy Reviews, 1995. **30**(3): p. 139-174.
6. Blanco, M. and M. Alcalá, *Content uniformity and tablet hardness testing of intact pharmaceutical tablets by near infrared spectroscopy: a contribution to process analytical technologies.* Analytica chimica acta, 2006. **557**(1-2): p. 353-359.
7. Brailean, J.C., et al., *Noise reduction filters for dynamic image sequences: A review.* Proceedings of the IEEE, 1995. **83**(9): p. 1272-1292.
8. Vaseghi, S.V., *Advanced digital signal processing and noise reduction.* 2008: John Wiley & Sons.
9. Xu, Y., et al., *Wavelet transform domain filters: a spatially selective noise filtration technique.* IEEE transactions on image processing, 1994. **3**(6): p. 747-758.

631244 The Use of Linear Regression and R^2 in Continuous Manufacturing to Dictate When Loss-in-Weight Data Is Acceptable to Control Feeding Following Feeder Refill

Abstract:

Continuous manufacturing is becoming an ever more popular approach to pharmaceutical manufacturing because it provides many advantages compared to conventional batch manufacturing. Continuous manufacturing utilizes feeders to constantly add material, at the appropriate concentration, to the system through rotating screws. This fed material will end up downstream in tablet or capsule form. One of the major advantages of continuous manufacturing is the ability to continuously monitor and control the amount of each material that is being fed into the system. This is accomplished by monitoring the weight loss in the feeders over time. The loss-in-weight data provides the feed rate (kg/hr) of the material. Maintaining the target feed rate of each material throughout production is critical to meeting the critical quality attributes of the product.

By monitoring the real time weight data the control system can make updates to the screw speed to ensure the target feed rate is being maintained. However, the feeder hoppers have a limited capacity so material needs to be added at regular intervals to refill the hoppers throughout the batch. While the refill is occurring, the system loses the ability to control and monitor the feed rate. Hence, it is imperative for the feeder to return to loss-in-weight feeding mode using real time weight data as quickly as possible.

A criteria is required to determine at what point the weight data coming into the system can be relied upon to control the feeders through loss-in-weight. The continuous manufacturing line at Thermo Fisher Scientific continuously monitors the weight loss using linear regression to determine the feed rate and how well the actual data matches the predicted data to provide continuously updated coefficient of determination (R^2) throughout the feeding process. While feeding in loss-in-weight mode the R^2 value remains close to 1. However, when in refill the value decreases to approximately zero. Once the refill has been added the slope of the weight loss begins to stabilize and the R^2 begins to increase. Thermo Fisher Scientific has implemented an approach that relies on the R^2 value obtained through linear regression of the weight loss to dictate when the loss-in-weight data coming into the system is once again acceptable to control the feed rate. Once, the R^2 value reaches its designated set point the system will change from volumetric to loss-in-weight feeding.

Studies will be presented examining different materials, feeder sizes, and refill methods. Feed rate, weight, screw speed, and R^2 data will be presented before, during, and after refills to show the behavior of the feeders throughout the entire process. The data will show how using linear regression with an R^2 set point standardizes the criteria to return to the desired state of loss-in-weight feeding following feeder refill while removing material and refill method variability inherent to the process.

631570 Continuous Manufacturing Control Strategy for Material Traceability

Abstract:

Material traceability plays an important role in pharmaceutical companies through root cause analysis, continuous improvement, value stream mapping, patient safety, regulatory compliances, and recalls. From a batch manufacturing perspective, the materials placed in the batch are easily known, as each batch is a separate set of material additions and operations. Consequently, if a non-conformance or recall event arise, the size of the lot to be placed in investigation or to be withdrawn from the market will be of the same order as its batch size.

However, when dealing with the same production in a continuous manufacturing line, the batch size is no longer tethered the equipment size or the material addition. Materials can be added to the continuous process many times during production. This can allow many lots of incoming raw material to be used during the process. In addition, the continuous process can be used to produce multiple batches of product without stopping or cleaning between each batch. This has the advantage of allowing the process to remain at steady state for longer, but many lots of incoming materials can be split amongst the produced batches. Tracking which batches have which lots of material is imperative.

This talk will detail a real-time method for determining lot locations within produced batches. This allows each batch's constituent ingredient lots to be determined even as the batch is completed. It also enables continuous production from one batch to the next. The results of experiments using tracer data from dispensing station to the output of the process will be presented, which shows the timing needed for a lot of material to move through the system. The strategy for utilizing this data for material tracking will be discussed.

626646 Development of a High-Fidelity Digital Twin Using DEM for Evaluating Continuous Manufacturing Control Approaches

Abstract:

Pharmaceutical process development (batch or continuous) includes multiple particle processing steps that could affect the drug product quality such as content uniformity. Such processes are normally developed by extensive experimental research, however, by narrow material variation and within a small design space. Thus, *in silico* modelling and high-fidelity digital twins can be beneficial for the industry to explore larger design space, including in-depth process development, and more optimization studies through numerous virtual experiments and by quantitative evaluation of the outcomes.

The Research Center Pharmaceutical Engineering (Austria) supported by the U.S. FDA have recently started a project to develop an approach for assessing drug manufacturing control strategies based on high-fidelity digital twins taking a direct compression line as a use-case. The high-fidelity digital twin platform is designed based on typical pharmaceutical materials, equipment, processes, and drug products. These components will be simulated in the digital twin tool by verified numerical methods and algorithms [1,2,3]. The model accuracy and simulation results will be validated by experimental methods to ensure the model predictability considering different direct compression lines.

This work presents the project workflow and the preliminary results. This includes the characterization of a common blend widely used in the pharmaceutical industry, introducing material properties for the blend in the Discrete Element Model (DEM), the workflow for calibrating the DEM contact parameters, and finally, the results of an investigation related to a study of a horizontal continuous mixer. The results of the mixer simulations cover design space studies such as different mixer throughputs and rotational speeds. Furthermore, materials are tracked in each unit operation by means of Residence Time Distribution (RTD), which can be used as a vital part of the control strategy. Combining RTD curves of individual units via convolution integrals, the line RTD curve can be calculated, which further describes propagation and attenuation of disturbances through a continuous manufacturing line for material traceability.

Disclaimer:

This abstract reflects the views of the authors and should not be constructed to represent FDA's views or policies.

References:

[1] Toson et al., 2018. "Detailed modeling and process design of an advanced continuous powder mixer". Int J Pharm 552, 288–300. https://doi.org/10.1016/j.ijpharm.2018.09.032

[2] Toson et. Al., 2019. "Explicit Residence Time Distribution of a Generalised Cascade of Continuous Stirred Tank Reactors for a Description of Short Recirculation Time (Bypassing)", Processes 2019, 7(9), 615; https://doi.org/10.3390/pr7090615

[3] Toson et al. 2021, "Continuous Powder Mixing Technology: Validation of the DEM Model".

submitted to J Pharm Sci.

629059 RTD Based Digital Twin of Continuous Pharmaceutical Manufacturing Process

Abstract:

Residence time distribution (RTD) is a probability distribution function that describes how long a fluid or powder element spends inside a given operation. Currently, continuous pharmaceutical manufacturing is facing several challenges such as prediction of mixing in each unit operation as well as in an integrated line, assurance of content uniformity (CU), and tracing of materials for lots and batches identification ideally needed for product regulation and urgent re-calls [1-3]. These issues are directly related with patient safety and therefore are critical to address. The RTD model has been found in the center of all solutions to address these critical issues. However, because of different level of complexities, the accurate development and effective applications of the RTD model is still a challenging task. Therefore, a systematic framework is needed through which, the RTD model can be developed and deployed for varies applications.

In this work, an RTD based digital twin has been developed for continuous pharmaceutical manufacturing process. The heart of the digital twin is a modular integrated flowsheet model capable of predicting the mixing in each unit operation as well as in an integrated line. The flowsheet model consists of the mathematical RTD models coupled with mechanistic models. The model library and knowledge base are two important supporting tools for the developed digital twin. The model library consists of the RTD model of basic elements such as plug flow process, perfectly mixed process or continuous stirred tank (CST), and dead zone (DZ). Subsequently, via combining these basic elements, the pharmaceutical unit operation (e.g. blender, feed frame) RTD model library has been developed. The knowledge base which has been developed through extensive experimentations consist of the RTD parameters for a broad range of equipment's and materials. The digital twin of any processes can be easily generated using these tools via drag and drop approach. The digital twin is flexible in nature, can be adapted for any continuous pharmaceutical manufacturing plant and thus should have a broad range of real time applications through distributed control system (DCS) as well as for offline research and development purposes.

The applications of the digital twin to predict the mixing capabilities and effect of disturbances have been demonstrated. The applications of the RTD based digital twin for real time diversion of tablets have been demonstrated to assure the content uniformity of the tablets. In CM, the drug concentration is measured in real time before the tablet compaction (chute &/or feed frame) using PAT sensor. The proposed control strategy then uses this inlet concentration to determine a signal for the diversion strategy that can accurately be used to reject tablets that are out of tolerance limits at the outlet of the tablet press. Two strategies, i.e. 'fixed window based strategy' and RTD based strategy have been developed, compared and evaluated. In fixed window approach, the tablet diversion is facilitated through knowledge of time delays from the point of detection to the point of the affect (tablet press outlet gate) in the system. The sensor that detects the concentration is connected to a comparator block which decides if the said concentration is within the specifications. If it is not within specification, the experimentally derived time delay is applied and post this the diversion begins. The diversion stops when a concentration within spec is detected and the another time delay is applied. In RTD based approach, the RTD is used to predict the outlet concentration from the inlet concentration. The predicted signal is then used to initiate the diversion. The first approach is simpler to implement but may lead to lower production efficiency. The second approach is based on more advanced RTD based technique and will ensure more efficiency but is relatively complex to implement. The applications of the RTD based digital twin for material traceability has been also demonstrated. A corresponding software prototype has been also developed to automate the material traceability procedure [3].

The objective of this presentation is to folds. First to highlight the development of the RTD based digital twin and then to demonstrate its applications for prediction of mixing, assurance of content uniformity via real time tablet diversion, and materials traceability.

References

[1). Singh, R. (2019). Systematic framework for implementation of RTD based control system into continuous pharmaceutical manufacturing pilot-plant. Pharma. Issue 34, 43-46.

[2]. Bhaskar, A., Singh, R. (2018). Residence time distribution (RTD) based control system for continuous pharmaceutical manufacturing process. Journal of Pharmaceutical Innovation. DOI: 10.1007/s12247-018-9356-7.

[3]. Billups, M., Singh, R. (2018). Systematic framework for implementation of material traceability into continuous pharmaceutical tablet manufacturing process. Journal of Pharmaceutical Innovation. https://doi.org/10.1007/s12247-018-9362-9.

623874 New Methods for Design Space Identification and a Re-Interpretation of ICH Endorsed Guide for Q8Q9Q10 Implementation

Abstract:

Since its inception, the concept of the design space has always been a model-based one [1]. One cannot establish a bridge between two experiments without the assumption of continuity. The tools and methods used to support this continuity determine the type of experiments required to build the continuum upon which a design space is to be built. This continuity is pragmatically defined with a mathematical model. Establishing a design space strictly through empirical means implies a heavy burden down the road as it pertains the verification and lifecycle of the design space. This is mostly driven by the need to verify both, the structure of the continuum (equations) and its parameters. The use of a mathematical formulation that is supported by fundamental principles of physics, chemistry and biology permutes the verification of the equation structure with the verification of the assumptions that sustained the choice of mathematical model during development. If the assumptions still hold in the system, the model must hold too. Furthermore, the verification of a probability postulation remains subject to the same stochasticity and offers inconsequential evidence given the number of at-scale experimentation a pharmaceutical company is realistically willing to do. This talk will present a review of recent methods to derive candidate design spaces and furthermore refine them into multivariate operational ranges to be used in master batch records in commercial manufacture [2,3,4]. A discussion will follow as to the logical ways towards verification and lifecycle of the multiple components that assemble the design space [5].

[1] Garcia-Munoz, S., Luciani, C.V., Vaidyaraman, S. and Seibert, K.D., 2015. Definition of design spaces using mechanistic models and geometric projections of probability maps. *Organic Process Research & Development*, *19*(8), pp.1012-1023.

[2] Laky, D., Xu, S., Rodriguez, J.S., Vaidyaraman, S., García Muñoz, S. and Laird, C., 2019. An optimization-based framework to define the probabilistic design space of pharmaceutical processes with model uncertainty. *Processes*, *7*(2), p.96.

[3] Kucherenko, S., Giamalakis, D., Shah, N. and García-Muñoz, S., 2020. Computationally efficient identification of probabilistic design spaces through application of metamodeling and adaptive sampling. *Computers & Chemical Engineering*, *132*, p.106608.

[4] Zhao, F., Grossmann, I.E., García-Muñoz, S. and Stamatis, S.D., 2020. Flexibility index of black-box models with parameter uncertainty through derivative-free optimization. *AIChE Journal*, p.e17189.

[5] ICH-Endorsed Guide for ICH Q8/Q9/Q10 Implementation, 2011, International Conference on Harmonisation of Technical Requirements for Registration of Pharmaceuticals for Human Use

632393 Molecular Simulations Unravel the Dynamics of Oiling out of β-Alanine

Abstract:

Crystallization is a purification process where a solute in a supersaturated environment undergoes self-assembly to form crystals. However, the degree of supersaturation may drive the solute molecules into a gel-like phase that coalesce to form crystals. In literature, this phenomenon is also known as oiling out or liquid-liquid phase separation (LLPS). The process of oiling out has been traditionally observed during the crystallization of organic molecules and biomolecules. More recently, such formation of gel-like phase has also been observed during the crystallization of Metal-organic frameworks (MOFs) as well. To obtain greater control over the process of crystallization and to obtain the desired outcome of crystallization, it is necessary to understand the molecular events which lead to oiling out of the solute molecule. In this work, we employ molecular simulations to simulate crystallization and oiling out of β-alanine molecules. Simulations uncover that supersaturation influences the configuration of solvent molecules around the solute and allows the formation of dense metastable phases before leading to crystallization. The interaction energy landscape obtained by analyzing various degrees of freedom shows that the minimum in the energy landscape lies out of the crystalline region during oiling out. Results suggest that kinetics, along with the energetics, influence the phenomenon of oiling out, and understanding solvation shell dynamics can help design better control strategies for crystallization.

623444 New Insights on the Effects of Molecular Rigidity and Purity on Crystallization Pathways

Abstract:

In recent years, there has been a steady increase both in molecular rigidity and molecular weight of substances with potential therapeutic benefits transiting from medicinal chemistry. It is believed that this increase had led to an increase of challenges to crystallization due in part to molecular rigidity and complexity[1]. In addition, increasing molecular weight is expected to result in increasing challenges to achieving acceptable chemical purity due to the increase in the number of related impurities and synthetic steps required to produce relatively large drug substances. In this talk, we present a case study of a drug candidate that displayed unique challenges to crystallization. Investigations revealed that molecular rigidity and impurities levels dictate the outcome of the crystallization attempts to either crystalline nucleation or oiling-out. An NMR-based approach is presented that proved critical in identifying operating conditions that steer the system toward crystallization. This approach, along with understanding the causes of oiling-out, allowed the development of a robust crystallization procedure that was critical for the progression of the drug substance into clinical trials.

1. Derdour L., A pathway to first crystals for substances prone to Liquid-Liquid phase separation, Chem. Eng. Technol., 44(3), 488–494, 2021

624919 Protein Crystallization with Gas Bubble Templates and Scaling-up Studies

Abstract:

Introduction

With growing interest in the protein drugs in the pharmaceutical company, crystallization of the protein becomes potential technology in downstream manufacturing purified products with advantages of stability, storage and delivery. Moreover, the crystallization process is easy to scale up with much lower manufacturing cost [1].

One of the useful method to facilitate crystallization process is to use heteronucleants to promote nucleation [2], such as nano-templates. However, concerns were raised due to generality of engineered liquid-solid interfaces in the larger industrial [3] by introducing potential impurities. Therefore, by using gas bubbles as heteronucleants is a potential method to accelerate crystallization process without introducing more impurities. Previous literatures have proven that the gas-liquid interface could promote the nucleation and achieve better control of crystal sizes for organic molecules [4–6].

This study investigated the effect of gas bubbles on the protein crystallization in a scaling-up methodology: air bubbles were started adding into the protein crystallization system in μL-level hanging-drop experiment; then different gas bubbles were mixed with crystallization solution by a mL-level microfluidic system; lastly gas flow was injected into a L-level batch crystallizer. In the small-scale system, the experiment containing gas bubble was compared to the one without gas bubble under the same conditions to gain insights who the presence of the gas-liquid interface affect the protein crystallization behavior. Microfluidic studies focused on how different gas feed altered the crystallization process. By using Process Analytic Tools (PATs), the crystallizer experiment gained more insights in the gas bubbles influence on the industrial scale-up of the protein crystallization.

Experimental Methods

Lysozyme in 0.1 M sodium acetate buffer with pH 4.2 was selected as the model crystallization system and sodium chloride with a range of concentrations were used to facilitate the crystallization process. The hanging-drop experiments were performed in 24-well plates sealed with glue to allow controlled vapor diffusion in closed systems as shown in Figure 1 (a), with a crystallization droplet of ~3 μL. For the 1 mL-scale study, a microfluidic chip was used to produce stable gas bubbles as shown in Figure 1 (c). The samples from both set-ups were stored at 20°C environment inside the thermostatic incubator and were taken out to be observed under the optical microscope. In the batch crystallizer of 500 mL illustrated in Figure 1 (e), the temperature was controlled by a heat exchanger covered the crystallizer and CryPRINS information system at 20°C and an array of PATs including FBRM, PVM, and UV/Vis probes was deployed to monitor the process characteristics.

Results and Implications

The results showed that the crystals were attached to the bubble. The crystals attached to the surface showed a curved facet as shown in Figure 1 (b) instead of the tetragonal geometry as in the bulk solution surface, confirming that crystal was actually growing on the gas-liquid interface. This piece of evidence also agreed with the heterogenous nucleation mechanism, where the air bubble provides the heterogenous nucleation site for lysozyme crystallization. To further understand the static air-bubble effect on crystallization, the crystal density near the injected air-bubble was compared to the one outside the air-bubble. The crystal density was increased up to 4 times due to the air-bubble in the range of the tested condition. As the gas-liquid interface provides a heterogenous nucleation sites, the free energy required for the lysozyme crystallization is lowered giving higher probability for lysozyme to nucleate near the air bubble.

In the microfluidic studies, four different types of gas, helium, oxygen, carbon dioxide and nitrogen, were injected in the microfluidic device with 100 mg/mL lysozyme and 1.1 M NaCl mixed solution. Figure 1 (d) depicts the gas bubble and crystal formation situation. For the experimented gas, it was found that the population density of lysozyme crystal showed a monotonical dependance on the gas solubility. The crystal tended to form greatest population density induced by carbon dioxide. Regarding the crystal formation locations, lysozyme crystals had a greater tendance to form on the gas-liquid interface, supporting a heterogeneous nucleation mechanism induced by the gas bubble. The solution treated with carbon dioxide contained the highest proportion, more than 80% of lysozyme formed on the carbon dioxide bubble surface. When gas bubbles were mixed into the batch protein crystallizer, the onset of nucleation was found to start faster, in agreement to the findings in the μL-scale hanging-drop study. The addition of gas bubbles is able to promote protein nucleation by acting as a heterogenous nucleation surface, paving the way for using the gas bubble as a soft template for protein crystallization in downstream processing.

Conclusions

Lysozyme crystals were discovered to be attached on the bubble surfaces. Air bubbles in the hanging-drop experiment accelerated the crystallization process, by reducing the induction time and increasing the population density in most of pH and lysozyme concentration conditions. Different gas bubbles in the microfluidic study shows the tendency of population density formed in the aqueous solution follows same as the one of the gas solubility. Meanwhile, CO_2 has the greatest possibility of protein crystals attached to the bubble surface. The promotion effect of gas bubbles were successfully applied in larger scale crystallization process, reducing the induction time and increase the yield. All the results demonstrated that the protein crystallization with air bubble as templates can be potential to accelerate protein crystallization in the industrial scale.

1. Yang, H., Chen, W., Peczulis, P. & Heng, J. Y. Y. Development and Workflow of a Continuous Protein Crystallization Process: A Case of Lysozyme. *Cryst. Growth Des.* **19**, 983–991 (2019).
2. Artusio, F. & Pisano, R. Surface-induced crystallization of pharmaceuticals and biopharmaceuticals: A review. *Int. J. Pharm.* **547**, 190–208 (2018).
3. Zhou, R. Bin, Cao, H. L., Zhang, C. Y. & Yin, D. C. *A review on recent advances for nucleants and nucleation in protein crystallization. CrystEngComm* **19**, (2017).
4. Wada, Y., Matsumoto, M. & Onoe, K. Antisolvent crystallization of NaCl using the minute-bubble technique – Effects of different antisolvent types. *J. Cryst. Growth* **448**, 76–81 (2016).
5. Kleetz, T., Pätzold, G., Schembecker, G. & Wohlgemuth, K. Gassing Crystallization at Different Scales: Potential to Control Nucleation and Product Properties. *Cryst. Growth Des.* **17**, 1028–1035 (2017).
6. Adhikari, B. M., Truong, T., Bansal, N. & Bhandari, B. Influence of gas addition on crystallisation behaviour of lactose from supersaturated solution. *Food Bioprod. Process.* **109**, 86–97 (2018).

Figure 1. (a) Schematic diagram of the hanging-drop experiment with gas-bubbles; (b) microscopic image of the crystal on bubble surface in the droplet; (c) schematic diagram of the microfluidic experiment with gas-bubbles; (d) microscopic image of the crystal formation inside capillary after mixing gas bubbles by microfluidic device; (e) schematic diagram of batch crystallizer fitted with gas bubble inflow and integrated PAT array; (f) PVM image of bubble and crystal formation inside the crystallizer.

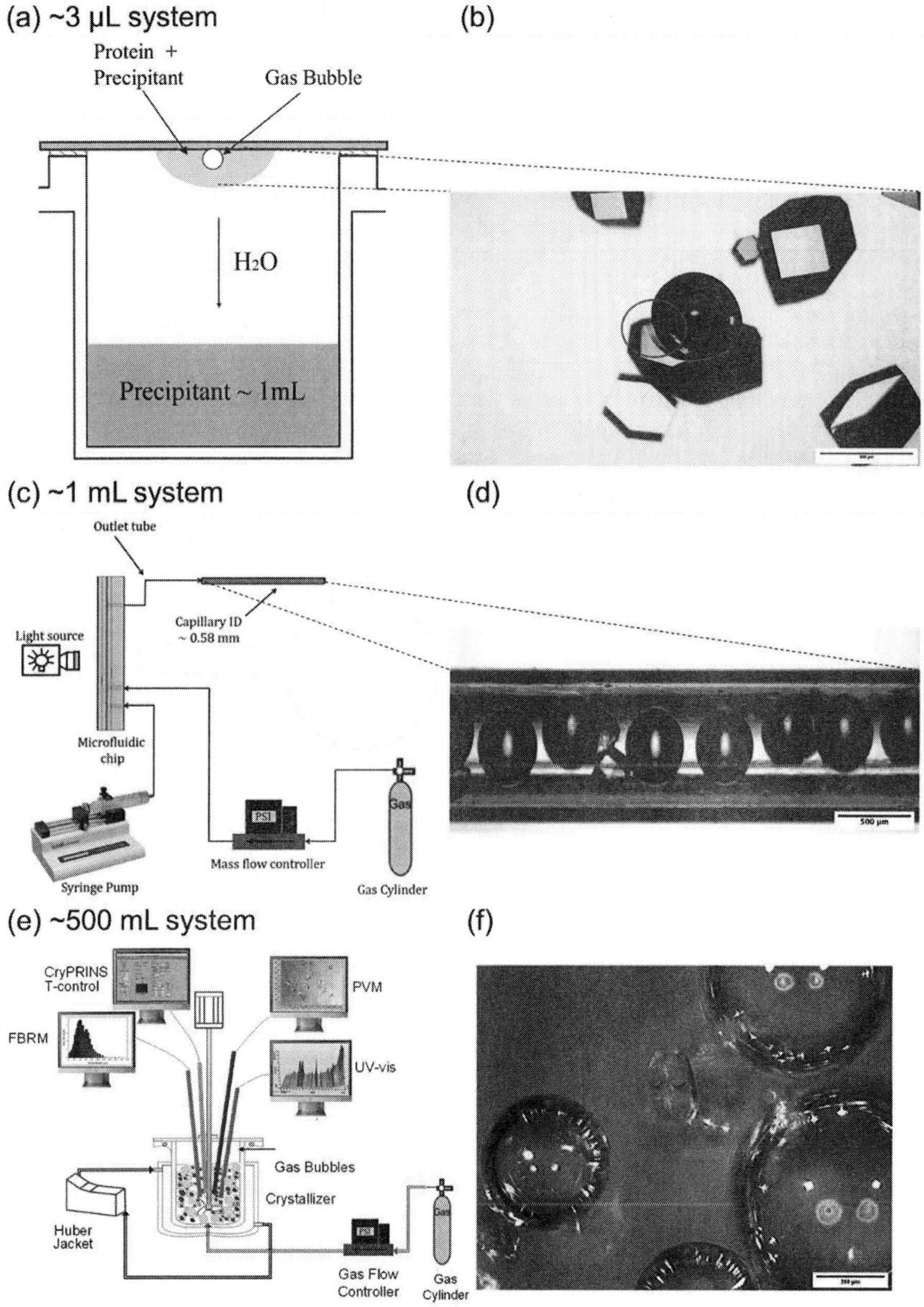
(a) ~3 μL system
Protein + Precipitant
Gas Bubble
H2O
Precipitant ~ 1mL
(b)
(c) ~1 mL system
Outlet tube
Light source
Capillary ID ~ 0.58 mm
Microfluidic chip
PSI
Gas
Mass flow controller
Syringe Pump
Gas Cylinder
(d)
500 μm
(e) ~500 mL system
CryPRINS T-control
PVM
FBRM
UV-vis
Gas Bubbles
Crystallizer
Huber Jacket
Gas Flow Controller
Gas Cylinder
(f)

627504 Crystallization Development for Molnupiravir, an Investigational Antiviral for the Treatment of COVID-19

Abstract:

Molnupiravir is a small molecule API (active pharmaceutical ingredient) pro-drug of a nucleoside analog which has shown potential for the treatment of COVID-19 in pre-clinical and clinical studies. Merck partnered with Ridgeback Biotherapeutics in May 2020 to accelerate development of the drug. Due to the high anticipated global demand, the team quickly worked on delivering a robust manufacturing route to reliably produce API of consistent quality. This talk will specifically focus on the development of a robust crystallization process which reliably delivers acceptable chemical purity and physical attributes, and is insensitive to equipment and scale of operation. The original isolation of API was achieved through a distillative crystallization, in which a solution of API, ethyl acetate, acetonitrile, and water is azeotropically dried to induce crystallization of the highly water-soluble API. The existing process was modified to give better control around the seeding point and initial de-supersaturation, which resulted in more consistent nucleation and growth, leading to more consistent API attributes across batches at different scales. To determine the level of control needed on API particle size for the capsule formulation, the impact of the high and low extremes of particle size ranges on bioavailability was assessed. Targeted experimentation also revealed how supersaturation controls the formation of the metastable and thermodynamically stable crystal forms. This presentation will discuss the methodology used to gain a thorough understanding of crystallization performance (morphology, particle size, bulk density, crystal form) over just a few months, ultimately producing API with consistent properties across various production sites in five different countries.

627758 An Experimental Study on the Ice Nucleation Properties of Aqueous Solutions in Vial Freezing Processes

Abstract:

Introduction

Freezing is an important process in the pharmaceutical and food industries to improve the stability and thus shelf life of various products. Ice nucleation is the first step of any freezing process; it is an activated process and occurs randomly in supercooled solutions. In small volumes, especially in the mL scale and below, supercooling in the order of 10 K and more is required to induce ice nucleation and thus freezing in a reasonable timeframe [1].

The prediction of the ice nucleation behavior of an arbitrary solution is challenging due to the stochasticity of ice nucleation and its interplay with solutes and heterogeneous nucleation sites. While there is a large body of research related to nucleation temperatures of microdroplets due to their importance in cloud formation [2], literature on larger volumes is sparse. In addition to homogeneous ice nucleation, the heterogeneous formation of ice nuclei is observed in industrial processes. Common heterogeneous nucleation sites range from suspended dust particles and dissolved gases to container surfaces. The effect of dissolved gases and particles especially is of importance in the pharmaceutical industry. Products frozen under GxP conditions, i.e. containing fewer impurities and thus heterogeneous nucleation sites, generally experience larger supercooling than those prepared under standard laboratory conditions. This difference in nucleation behavior between laboratory and GxP-conform production scale proves to be a major hurdle in upscaling of lyophilization processes and is thus of great practical relevance [3].

We thus aimed to develop a general understanding of the nucleation behavior of arbitrary aqueous solutions in GxP and non-GxP settings. To do so, we have carried out an experimental campaign in a batch crystallizer at 1 mL scale, close to the conditions in industrial processes. Our work focuses on quantifying the effect of solutes and setup-dependent heterogeneous nucleation sites to gain insight that can be used for the design of freezing processes.

Methods

The ice nucleation in a temperature-controlled vial is assumed to be an inhomogeneous Poisson process and we have estimated the nucleation rate J for both water and aqueous solutions in 1 mL batches [4-5]. To generate datasets of statistically meaningful size for this parameter estimation, we have carried out freeze-thawing experiments, i.e. we have been freezing and consequently thawing the same solutions several times in sequence. 10-50 of such cycles in 15 vials in parallel, generating in total 150-750 nucleation events, have been carried out per experimental run. We describe the effects of heterogeneity and solutes on the ice nucleation behavior of water with the water activity-based ice nucleation model (WABINT) [2]. WABINT assumes that the nucleation rate depends only on water activity and the type of nucleation sites present; the effect of solute thus is assumed to be limited to its influence on water activity and independent of the type of solute. We have analyzed solutions containing different types of solutes to investigate whether this assumption holds for the systems of interest.

Results

We have performed experimental studies in the batch crystallization system on water and aqueous solutions of sucrose using various cleaning protocols.

First, we have frozen 15 vials containing ultrapure water under standard laboratory settings to investigate the intrinsic stochasticity of ice nucleation and to validate the experimental setup. Fig. 1a and 1b provide an overview over the temperature profiles used and show the obtained outcome of a single cycle for one cooling block with four vials. It is evident, that the nucleation events affect each other: The nucleation of vial A1 leads to an increase in temperature of all other vials in the same cooling block. This is due to the heat release upon nucleation and it is inherent to the setup. Moreover, we have conducted additional studies with only one vial in each cooling block to investigate the extent to which such thermal interaction leads to additional broadening of the distributions.

We have confirmed that the presence of thermocouples inside the vials did not affect the nucleation behavior of the freezing water for non-GxP conditions. Since literature on lyophilization in GxP settings commonly reports thermocouple effects on ice nucleation, we have developed an indirect approach for temperature measurement to run nucleation experiments with GxP vials without the need of opening the vials and to insert thermocouples.

Fig. 1c provides the experimental nucleation temperatures for the freeze-thawing of ultrapure water. We have observed not only the inherent stochasticity of nucleation for each individual vial, but also significant inter-vial variability. Some vials experienced very narrow temperature distributions with a broadness of less than 1 K; this is in accordance with literature studies on the freezing of microdroplets [6]. However, some other vials experienced broader distributions or showed a time-dependent decrease in nucleation temperature such as vial A1. This variability in nucleation behavior was observed despite extensive cleaning steps covering all parts of the experimental setup.

The source of such vial-to-vial variability is not well understood although it is widely reported; experimental studies on freezing in lyophilization report nucleation temperature distributions with broadness of up to 20 K for a single batch of vials containing the same product. We hypothesize that such variability is the consequence of both a variability in heterogeneous nucleation sites and the thermal interaction between vials. We have carried out an experimental campaign to provide insight into this phenomenon.

Second, fig. 1d shows the outcome of a similar experiment with 20 wt.% sucrose solutions. We observed a depression of the average nucleation temperature from -8.8°C for water to -11.8°C for the sucrose samples; this is in line with literature on nucleation temperature depression. The depression, however, did not lead to overall lower nucleation temperatures in all vials, but primarily consisted in suppressing nucleation events at high temperature, which we quantify based on a reduction in standard deviation of the nucleation temperature from 3.4 K to 1.2K. It is reasonable to conclude that the addition of solute affects the individual sources of heterogeneous nucleation differently, which is a well-known effect in literature in smaller systems [7], but was not yet studied at larger scale. We thus have carried out further experiments to investigate this effect.

Conclusion

We present for the first time an experimental study on ice nucleation in a parallelized batch crystallizer at 1 mL scale. Our experimental data confirmed the presence of a vial-to-vial variability in addition to the inherent stochasticity of ice nucleation and of complex solute effects. Our experimental campaign includes the investigation of systems with various types of solutes and solutions prepared under different levels of cleanliness to improve our understanding of the ice nucleation process in these systems. Such understanding is crucial for the design and development of processes such as lyophilization and freezing for the purposes of storage, where a homogeneous freezing step is essential.

Acknowledgement

The authors thank The Janssen Pharmaceutical Companies of Johnson & Johnson for the support in the course of this project.

References

[1] J. Searles, J. Carpenter and T. Randolph: The ice nucleation temperature determines the primary drying rate of lyophilization for samples frozen on a temperature-controlled shelf, *Journal of Pharmaceutical*

Sciences (2001), 90, 7.

[2] B. Zobrist, C. Marcolli, T. Peter, and T. Koop: Heterogeneous Ice Nucleation in Aqueous Solutions: the Role of Water Activity, *J. Phys. Chem. A* (2008),112,3965-3975.

[3] S. Nail et al.: Recommended Best Practices for Process Monitoring Instrumentation in Pharmaceutical Freeze Drying, *AAPS PharmSciTech* (2017), 18, 7.

[4] G. M. Maggioni and M. Mazzotti: Modelling the stochastic behaviour of primary nucleation, *Faraday Discussions*, 179, 359 (2015).

[5] G. M. Maggioni, L. Bosetti, E. dos Santos and M. Mazzotti: Statistical Analysis of Series of Detection Time Measurements for the Estimation of Nucleation Rates, *Cryst. Growth Des.* (2017), 17, 5488–5498.

[6] P. Wilson and A. Haymet: The Spread of Nucleation Temperatures of a Sample of Supercooled Liquid Is Independent of the Average Nucleation Temperature, *J. Phys. Chem. B* (2012), 116, 45, 13472–13475.

[7] S. Özilgen and D. Reid: The Use of DSC to Study the Effects of Solutes on Heterogeneous Ice Nucleation Kinetics in Model Food Emulsions, *LWT – Food Science and Technology* (1993), 26, 2, 116-120.

626783 A Novel Technology for Continuous Vacuum Drying of an API Slurry: Characterization of Product

Abstract:

Currently used drying technologies are mostly batch operations. After crystallization the API needs to be filtered and washed followed by drying. Mostly, these operations are carried out in batch, also because only very few types of equipment are available that would allow continuous operation of those processes. In recent years efforts have been made to develop and optimize continuous processes for pharmaceutical production. This development is additionally driven by the need to have fast, adaptable process equipment available. The advantage of continuous operation has been reported and discussed lately and it is important to understand that not only cost of production and advantages in scaleup drive this development.

Drying is a complex process during which various processes influence the final product properties. The properties of particles and crystals are usually well engineered during crystallization and should not be changed during the drying and filtration step. However, often agglomeration as well as attrition are observed, especially in filter bed drying. Additionally, many problems occur in this respect when considering scale up and transfer to another production site. Nonetheless this technology is often preferred because it has better drying efficiency than plate dryers because some movement is applied to the powder bed. Still this drying process can take many hours and very often yields a product that is not homogeneous and needs to be milled and sieved after production.

Part of the problem with agglomeration is the "sticky phase". During this phase of drying the moisture content of the powder bed provides the ideal composition for agglomeration. Applied shear during this phase is crucial because too little would result in the bed hardening and too much could cause very large or hard agglomerates to be formed. Continuous technologies have the advantage that the residence time of the process is much shorter (range of minutes) combined with the fact that exposure to process conditions is more homogeneous in this intensified process.

In this work we use a different type of technology that enables continuous drying of the API slurry in industrial scale. As a test substance we used Ibuprofen, which has needle-shaped particles which are prone to attrition as well as agglomeration. To simulate the slurry, we mix the ibuprofen with a mixture of 95w%water and 5% ethanol in a ratio of 1:1 to generate a pump-able slurry. In this ratio the ethanol simulates the residual solvent in which the solubility of the substance is very high and therefore harder agglomerates could be formed. The slurry is fed with a feed rate between 1-2kg/h to the dryer via a specially developed feeding tank that enables air free undisturbed feeding of the highly viscous slurry to the dryer.

In the presentation we show the results of a DoE that was carried out to investigate the residual moisture of the API after drying in one step from slurry to below 1% residual moisture. We will show the influence on the particle size and how the "sticky phase" can be overcome without changing the initial particle size and shape.

627717 Use of Bead Mixtures As an Enabling Approach for Optimal Production of Drug Nanosuspensions

Abstract:

Preparation of drug nanoparticles is one of the most popular approaches in dissolution enhancement of poorly water-soluble drugs [1,2]. Due to their higher surface area and mass transfer coefficient, drug nanoparticles dissolve faster than their micron-sized counterparts [3]. In addition, sub-100 nm particles have higher saturation solubility owing to their high curvature [3]. Wet stirred media milling (WSMM) has been the most preferred method for the manufacture and preparation of marketed drug nanocrystal-based products and nanocrystal suspensions in scientific studies according to a recent literature review [2]. The reason for this high demand in WSMM is that it has the potential of producing high drug-loaded, stable nanosuspensions while being a reproducible, scalable, solvent-free, and environmentally benign process [4–6]. Most of the WSMM studies in the literature focuses on the formulation aspects such as stabilization of the milled suspensions against aggregation. In contrast, there have been fewer studies focusing on the processing–operational challenges such as cost due to high energy consumption, long operating hours, high cooling demand due to heat dissipation, and contamination of drug particles by the beads [2,4,7,8]. In order to address the operational challenges, a mechanistic understanding of the impact of process parameters such as bead concentration, bead material, and stirrer speed should be developed.

Unlike any study in the WSMM literature including non-pharmaceutical fields, as a major novelty, this study aims to explore the feasibility of cross-linked polystyrene (CPS)–yttrium-stabilized zirconia (YSZ) bead mixtures as a novel optimization approach for fast, efficient production of drug nanosuspensions via WSMM. While both CPS and YSZ beads were utilized in prior WSMM studies [9,10], they have not been used in a bead mixture before. We hypothesize that bead mixtures allow for optimal milling of drug suspensions from a combined energy–cycle time–heat dissipation perspective. In this study, fenofibrate (FNB) was selected as a poorly water-soluble model drug. A total of 20 experiments were conducted at two stirrer speeds (ω = 3000 and ω = 4000 rpm) and two volumetric fractional bead loadings (c = 0.35 and c= 0.50) with five CPS–YSZ bead mixtures including CPS alone and YSZ alone (CPS:YSZ 0:1–1:0 v/v). About 235 g pre-suspensions that have 10% FNB, 7.5% HPC-L (L grade of hydroxypropylcellulose), and 0.05% SDS (sodium dodecyl sulfate) with respect to 200 g deionized (DI) water were prepared, wherein HPC-L and SDS served as the stabilizers based on our prior work [11,12]. XRPD was used to assess the crystallinity of the drug, while viscometry was used to determine the dynamic viscosity and examine the shear-thinning behavior of the milled suspensions. The timewise evolution of the FNB particle size was investigated via laser diffraction by taking suspension samples at various milling times. To characterize the breakage kinetics, three different kinetic measures were defined and analyzed including the apparent breakage rate constant of an nth-order kinetic model. A microhydrodynamic model was used to elucidate the different microhydrodynamic behavior of the CPS vs. YSZ beads and their pros/cons [13,14]. Various optimization criteria based on industrial considerations were adopted and relevant merit scores were calculated. Kinetic parameters and specific power–energy consumption were used together to assess the impact of bead mixtures.

Fig. 1 (left panel) presents the average power per unit volume P_w and time required for the median particle size to reach 0.25 µm t_{d50} as a function of the YSZ bead concentration in the mixture c_{YSZ} for four different operating conditions. As a general trend, higher c_{YSZ} led to higher lower t_{d50} at higher P_w. In other words, CPS beads were found to be efficient in terms of power consumption but YSZ beads provide faster breakage, which results in shorter cycle time to achieve the desired particle size. The microhydrodynamic model attributed the faster breakage with the YSZ beads compared to the CPS beads to their more frequent, energetic/forceful collisions than those of CPS beads owing to the higher density–elastic modulus of the YSZ beads.

An ideal WSMM process should produce a targeted drug particle size at the shortest cycle time with the lowest energy expenditure without serious heat dissipation (temperature control) issues. Hence, we defined a merit score, as shown at the top right panel of Fig.1, to guide our optimal process selection. We considered three major factors: energy expenditure (cost), cycle time, and heat dissipation related issues, and created four different scenarios with different values of the weighting coefficients w_i in the merit score definition. A higher merit score indicates that the operating parameters and bead mixtures selected are more desirable within the context of specific process optimization criteria considered. There are four scenarios: one wherein all three factors contribute to the merit score equally ($w_1 = w_2 = w_3 = 1$), a scenario wherein energy expenditure is considered less significant compared to cycle time and temperature related process shutdowns (most likely scenario in pharmaceutical industry as compared with other industries, reduced w_2: $w_2 = 0.1$), a scenario wherein heat dissipation related process shutdowns as related to P_w are partly mitigated by a high-capacity chiller (reduced w_1, $w_1 = 0.1$), and a scenario wherein milling cycle time is considered less important than other two factors (reduced w_3, $w_3 = 0.1$). In all scenarios (not shown here for brevity), 3000 rpm and 50% loading were optimal (highest merit scores). Moreover, in 9 out of 16 total cases examined (4 different processes with various ω–c in 4 different scenarios), a bead mixture had the highest merit score as opposed to CPS or YSZ beads alone. Fig. 1 illustrates that in the most likely scenario relevant to the pharmaceutical industry ($w_1 = w_3 = 1$, $w_2 = 0.1$), the highest merit score corresponded to the following milling conditions: 3000 rpm and 50% loading of 0.5:0.5 v/v CPS:YSZ.

Overall, this study has demonstrated that mixtures of two different bead types, YSZ and CPS, have achieved optimal wet stirred media milling (WSMM) when specific energy consumption (cost), milling cycle time (breakage kinetics), and heat dissipation related process shutdowns are simultaneously considered. This comprehensive, combined experimental–modeling study has revealed various insights as to why bead mixtures vs. YSZ or CPS beads alone offer much needed flexibility for WSMM optimization. Future work will focus on heat transfer analysis of the milling process, examination of the impact of bead mixtures on media wear and contamination and development of a new microhydrodynamic theory for bead mixtures.

REFERENCES

1. Kesisoglou F, Panmai S, Wu Y. Nanosizing — Oral formulation development and biopharmaceutical evaluation. Adv Drug Del Rev. 2007;59(7):631-44. doi: https://doi.org/10.1016/j.addr.2007.05.003.

2. Bhakay A, Rahman M, Dave RN, Bilgili E. Bioavailability Enhancement of Poorly Water-Soluble Drugs via Nanocomposites: Formulation‑Processing Aspects and Challenges. Pharmaceutics. 2018;10(3):86. doi: 10.3390/pharmaceutics10030086.

3. Bhakay A, Merwade M, Bilgili E, Dave RN. Novel aspects of wet milling for the production of microsuspensions and nanosuspensions of poorly water-soluble drugs. Drug Dev Ind Pharm 2011;37(8):963-76. doi: 10.3109/03639045.2010.551775.

4. Li M, Azad M, Davé R, Bilgili E. Nanomilling of Drugs for Bioavailability Enhancement: A Holistic Formulation-Process Perspective. Pharmaceutics. 2016 doi: 10.3390/pharmaceutics8020017.

5. Peltonen L. Design Space and QbD Approach for Production of Drug Nanocrystals by Wet Media Milling Techniques. Pharmaceutics. 2018;10(3):104. doi: 10.3390/pharmaceutics10030104.

6. Li M, Yaragudi N, Afolabi A, Dave R, Bilgili E. Sub-100nm drug particle suspensions prepared via wet milling with low bead contamination through novel process intensification. Chem Eng Sci. 2015;130:207-20. doi: https://doi.org/10.1016/j.ces.2015.03.020.

7. Bilgili E, Guner G. Mechanistic Modeling of Wet Stirred Media Milling for Production of Drug Nanosuspensions. AAPS PharmSciTech. 2020;22(1):2. doi: 10.1208/s12249-020-01876-w.

8. Kawatra SK. Advances in comminution. Littleton, Colorado: Society for Mining, Metallurgy, and Exploration; 2006.

9. Parker N, Rahman M, Bilgili E. Impact of media material and process parameters on breakage kinetics–energy consumption during wet media milling of drugs. Eur J Pharm Biopharm. 2020;153:52-67. doi: https://doi.org/10.1016/j.ejpb.2020.05.024.

10. Nakach M, Authelin J-R, Perrin M-A, Lakkireddy HR. Comparison of high pressure homogenization and stirred bead milling for the production of nano-crystalline suspensions. Int J Pharm. 2018;547(1):61-71. doi:https://doi.org/10.1016/j.ijpharm.2018.05.042.

11. Azad M, Afolabi A, Bhakay A, Leonardi J, Davé R, Bilgili E. Enhanced physical stabilization of fenofibrate nanosuspensions via wet co-milling with a superdisintegrant and an adsorbing polymer. Eur J Pharm Biopharm. 2015;94:372-85. doi: https://doi.org/10.1016/j.ejpb.2015.05.028.

12. Knieke C, Azad MA, Davé RN, Bilgili E. A study of the physical stability of wet media-milled fenofibrate suspensions using dynamic equilibrium curves. Chem Eng Res Des. 2013;91(7):1245-58. doi: https://doi.org/10.1016/j.cherd.2013.02.008.

13. Parker N, Rahman M, Bilgili E. Impact of media material and process parameters on breakage kinetics–energy consumption during wet media milling of drugs. Eur J Pharm Biopharm. 2020;153:52-67. doi: https://doi.org/10.1016/j.ejpb.2020.05.024.

14. Eskin D, Zhupanska O, Hamey R, et al. Microhydrodynamic analysis of nanogrinding in stirred media mills. AIChE J. 2005;51(5):1346-1358, doi: https://doi.org/10.1002/aic.10392.

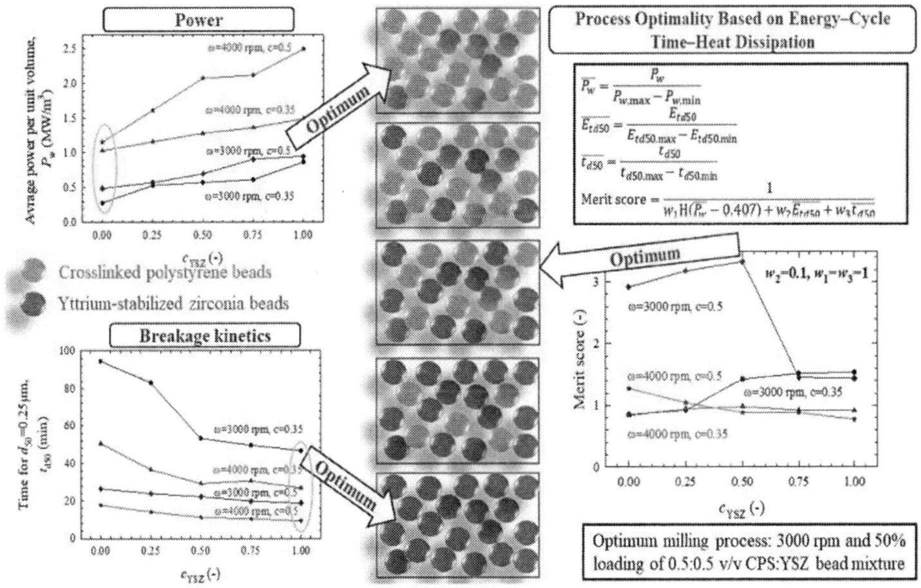

Fig. 1. Schematic summary of the impact of bead mixtures on power consumption, breakage kinetic parameter, and the overall merit score.

622859 A Robust Micro-Dosing Approach for Pharmaceutical Powders Based on a Small-Scale Powder Pump Concept

Abstract:

Despite recent developments in the field of micro-dosing of pharmaceutical powders, the exact dosing of cohesive powders remains challenging. Either the dosing equipment needs to be adjusted to the flow characteristics of the powder or the range of powder is limited to specially engineered powders exhibiting favorable behavior. Choosing the right option requires extensive investigation on the powder characteristics and adjustments of the feeding device based on experience. A promising technique showing low sensitivity to the powder characteristics is a micro-feeder based on a powder pump concept [1–3]. While this technique is originally applied for providing low feed rates of pharmaceutical powders to continuous processes[1–3], its robustness is also promising in the field of micro-dosing. A fast and robust solution for dosing small amounts of powder precisely, i.e., 10 to 500 mg, could speed up formulation development and early-stage manufacturing, and enable the production of personalized medication.

Investigating if this powder pump concept is capable to fulfill these requirements, a device was developed combining the feeding device itself, and a weighing cell to allow gravimetric dosing. A control strategy was developed increasing dosing precision by continuous monitoring of the dispensed volume and the dosed mass. Based on these quantities the bulk density in the relevant volume of the powder bed is calculated iteratively allowing a precise estimation of the volume that needs to be dispensed. The performance of the built device was assessed in experiments with different pharmaceutical powders, i.e. lactose monohydrate (Flowlac 90, Capsulac 60, Sorbolac 400), microcrystalline cellulose (Emcocel 90M, Vivapur 12, 101, 105, 200), magnesium stearate, and Ibuprofen 25. Furthermore, the influence of process parameters such as the speed of the pump piston, the scraper speed, and the number of partial doses on the dosing precision was studied.

The talk will highlight the influence of cohesion on the dosing precision and overcoming limitations regarding this influence by mechanical adjustments to the device, tuning of the process parameters, and the implemented control strategy. Based on the experimental data the benefits and limitations of this dosing concept, compared to state of the art equipment, will be discussed.

Literature Cited

1. Besenhard MO, Fathollahi S, Siegmann E, Slama E, Faulhammer E, Khinast JG. Micro-feeding and dosing of powders via a small-scale powder pump. *International journal of pharmaceutics.* 2017;519(1-2):314–322.
2. Fathollahi S, Sacher S, Escotet-Espinoza MS, DiNunzio J, Khinast JG. Performance Evaluation of a High-Precision Low-Dose Powder Feeder. *AAPS PharmSciTech.* 2020;21(8):301.
3. Sacher S, Heindl N, Afonso Urich JA, Kruisz J, Khinast JG. A solution for low-dose feeding in continuous pharmaceutical processes. *International journal of pharmaceutics.* 2020;591:119969.

626792 Pharmaceutical HME Process Development: Understanding the Impact of Kneading Elements in the Screw Configuration and Product Quality

Abstract:

Introduction

Hot melt extrusion (HME) is a continuous manufacturing process primarily carried out using co-rotating intermeshing twin-screw extruders (TSE). In the pharmaceutical context, HME is primarily used to produce amorphous solid dispersions of poorly soluble active pharmaceutical ingredients (APIs). In addition, it can also facilitate the development and production of products with crystalline API embedded into a polymer matrix and nanopharmaceuticals. Being a continuous manufacturing technology, HME based drug development typically requires prohibitively high amounts of API (kilograms of premix vs grams available in the early development stages) for successful formulation screening and early process development. The process setup itself is modular, allowing the process to be tailor-made for any formulation. This might however be problematic in the context of choosing the appropriate and efficient setup for an unknown formulation and it is an additional reason for relatively high quantities of material usually needed for the screening. Effectively solving the multidisciplinary challenge of formulation development, early process screening, effective scale-up and transfer to GMP production represents one the key challenges of the pharmaceutical industry and is especially challenging for novel technologies like HME.

Methodology

Addressing these challenges, our group has worked on the development of *in silico* and experimental tools for simpler process development and scale-up. Since to this date most of the process setup and scale-up activities are performed experimentally and empirically, one of the goals of our group was to create *in silico* tools for a rational, science-based process setup and scale-up, while addressing other important aspects, such as API degradation and overall product quality [1], [2]. The fundamental idea behind process modelling is the break down and the detailed analysis of the key process aspects, among which the most prominent might be the analysis of flow patterns developed as a result of the rotation and geometry of the individual screw element pairs [1], [3]–[8]. Following our work showing the impact an individual kneading section along the screw configuration has on the product quality in [9], [10], our work focused on detailed analysis of kneading elements and understanding their impact on the melt flow. For this purpose, we performed a number of Smoothed Particle Hydrodynamics (SPH) simulations of different kneading elements typically used in the pharmaceutical HME. Kneading elements with an angle of 30°, 45°, 60° and 90° between the individual kneading discs were investigated, with kneading block thicknesses of 1.1 mm, 2.55 mm and 4.1 mm, Figure 1.

Figure 1. Example of a 30° kneading element with three different disc thicknesses (1.1 mm, 2.55 mm and 4.1 mm).

Results

The kneading elements were analyzed in terms of their pressure and power characteristics, mixing efficiency, shear rate distribution and residence time. The data collected from the pressure and power characteristics shows the impact the kneading element angle and disc thickness have on the ability of the screw to convey melt, build pressure and dissipate energy. The differences found have an impact on the use of the individual kneading elements in the full screw configuration and processing tools. In addition, the analysis of the shear rate distribution and residence time distribution made it possible to produce reduced order models of those quantities that can be used in simplified 1D models of the process. This allows us to move from simplified equations representing the process state in 1D process models and directly use the data generated from detailed 3D screw simulations. In the light of the work presented in [9] and [10], this will help in developing a fully *in silico* based process development and scale-up framework that will be able to accurately calculate the process state and predict the resulting product quality. This will help to significantly reduce the time to market, cost and waste produced in the development of novel drug forms.

Literature

[1] J. Matić, A. Witschnigg, M. Zagler, S. Eder, and J. Khinast, "A novel in silico scale-up approach for hot melt extrusion processes," *Chem. Eng. Sci.*, vol. 204, pp. 257–269, Aug. 2019.

[2] J. Matić, A. Paudel, H. Bauer, R. A. L. Garcia, K. Biedrzycka, and J. G. Khinast, "Developing HME-Based Drug Products Using Emerging Science: a Fast-Track Roadmap from Concept to Clinical Batch," *AAPS PharmSciTech*, vol. 21, no. 5, p. 176, Jul. 2020.

[3] A. Eitzlmayr and J. G. Khinast, "Co-rotating twin-screw extruders: Detailed analysis of conveying elements based on smoothed particle hydrodynamics. Part 1: Hydrodynamics," *Chem. Eng. Sci.*, vol. 134, pp. 861–879, Sep. 2015.

[4] A. Eitzlmayr and J. G. Khinast, "Co-rotating twin-screw extruders: Detailed analysis of conveying elements based on smoothed particle hydrodynamics. Part 1: Hydrodynamics," *Chem. Eng. Sci.*, vol. 134, pp. 861–879, Sep. 2015.

[5] A. Eitzlmayr, J. Matić, and J. G. Khinast, "Analysis of flow and mixing in screw elements of corotating twin-screw extruders via SPH," *AIChE J.*, vol. 63, no. 6, pp. 2451–2463, Jun. 2017.

[6] A. Eitzlmayr *et al.*, "Experimental characterization and modeling of twin-screw extruder elements for pharmaceutical hot melt extrusion," *AIChE J.*, vol. 59, no. 11, pp. 4440–4450, Nov. 2013.

[7] A. Eitzlmayr *et al.*, "Mechanistic modeling of modular co-rotating twin-screw extruders," *Int. J. Pharm.*, vol. 474, no. 1–2, pp. 157–176, Oct. 2014.

[8] R. Baumgartner, J. Matić, S. Schrank, S. Laske, J. Khinast, and E. Roblegg, "NANEX: Process design and optimization," *Int. J. Pharm.*, vol. 506, no. 1–2, pp. 35–45, Jun. 2016.

[9] J. Matić *et al.*, "Towards predicting the product quality in hot-melt extrusion: Small scale extrusion," *Int. J. Pharm. X*, vol. 2, p. 100062, Dec. 2020.

[10] J. Matić *et al.*, "Towards predicting the product quality in hot-melt extrusion: Pilot plant scale extrusion," *Submitted.*

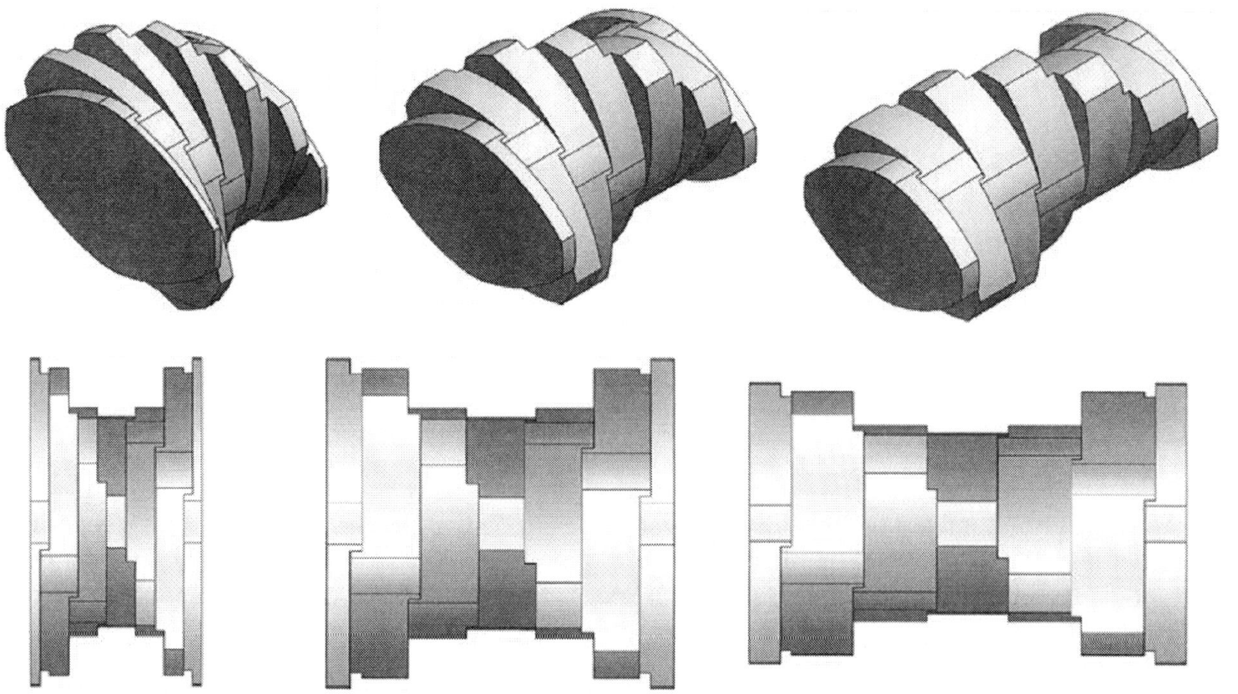

622877 Continuous Low Dose Powder Feeder

Abstract:

Continuous feeding of small quantities of powder is a challenge in a wide range of processes and industries. In pharmaceutical industry, highly potent active pharmaceutical ingredients (APIs) and very low density excipients e.g., silicon dioxide, are notoriously hard to feed with currently available commercial technology. The novel micro-feeder system is capable of feeding powders with different particle sizes and flow properties in low dose range of 1-80 g/h. It is based on active volumetric displacement. For feed rate determination the feeder is combined with a high precision balance at the outlet.

The talk will discuss opportunities and challenges of the novel device for continuous feeding of various powder bulk materials. In addition, approaches to overcome variation of displacement density in powder bed due to wall friction or relaxation are presented. A case study shows the integration in a continuous manufacturing line. The results indicate high reproducibility and accuracy of the system and its high potential to support continuous manufacturing lines consistently with low amounts of powders.

623009 Analysis of Transformed Bioprocess Data and Application of Hybrid Modeling Approaches to Enhance Upstream Process Development

Abstract:

Application of data-driven approaches in upstream process development has become a focal point of increased interest in biopharmaceutical industry. This, in part, is due to an increasing volume of on-line and sample data that become readily available during the process development, and upstream bioreactors being the key steps for determining the maximum achievable product quality and quantity. Despite significant advances made in the area (e.g., application of Multivariate Analysis), two key challenges remain: how do the intentional changes made during the process (e.g., feeding) impact the process and cell biology, and how do we use such knowledge to develop more productive processes? To address these questions, this presentation showcases usage of transformed bioprocess data and application of hybrid modeling approach (i.e., combination of data and fundamental-derived knowledge).

623356 Challenges in the Pharmaceutical Development of Lipid Nanoparticle Therapeutics

Abstract:

The discovery of RNA therapeutics has emerged as a promising modality to treat infectious diseases, cancer [1] and undruggable targets [2]. The delivery of RNA is a challenge because these molecules quickly get cleared upon administration and their size, along with negative charge does not allow it to cross into the cellular membrane [1]. As a result, many other platforms including positively charged antibody conjugates, masked endosomolytic agent-dynamic polyconjugates, or lipid nanoparticles offer an oligonucleotide delivery. These platforms also allow for DNA and RNA gene therapy, and vaccines. A recent rise in the development of clinical therapeutic RNA has led to the approvals of siRNA drugs (Patisiran) and mRNA vaccines (Moderna, CureVac, BioNTech) using lipid nanoparticle delivery [1].

The lipid nanoparticle delivery method offers the most mature platform for RNA therapeutics and has gained significant attention over the recent years [1]. However, there remains significant challenges in order to advance to commercialization. One of these challenges, requires efficient delivery of the RNA cargo to the targeted tissue and cellular compartment. Furthermore, the control of the lipid nanoparticle assembly is critical to the understanding the influence of in vivo pharmaceutical properties. A thorough comprehension of the physio-chemical properties to its bio performance is essential to ensuring the best quality commercialized drug product. This includes the molecular size, charge and polarity. The lipid nanoparticle assembly mechanism is also largely unknown [3]. We present an experimental design study of RNA loaded lipid nanoparticle self-assembly via a rapid antisolvent precipitation process, to identify critical parameters in the formation of LNPs. This will also help us understand the mechanism of the LNP assembly driving force to increase robustness for future formulation development and improve reproducibility.

RNA loaded lipid nanoparticles are complex structures of approximately 80-100nm, composed of cationic and neutral lipids. This includes PEGylated lipids in order to improve steric stability. The cationic lipid enables nucleic acid encapsulation and endosomal escape into the cytosol. The neutral lipids include phospholipid and cholesterol, which modulate the physical properties including rigidity, stability and biological membrane interactions. The oligonucleotide cargo, or RNA, is encapsulated at a defined ratio of cationic lipid to oligonucleotide. The process of forming the LNPs utilizes a micromixing device to generate controlled nanometer size, narrow polydispersity, and high RNA encapsulation efficiency [3]. The design of the mixing device is a continuous flow operation allowing for a rapid precipitation process suitable for commercia scale-up [3].

The goal of our developmental strategy is to control the LNP assembly process and reproducibility and correlate physicochemical properties with efficacy and/or toxicity. We use several analytical tools to enable robust and sensitive drug product characterization. For example, small-angle X-ray scattering (SAXS) quantitatively characterizes the structural features of RNA loaded LNPs [3]. Additionally, cryogenic transmission electron microscopy (cryo-TEM) micrographs can be used to identify LNP morphology, either hexagonal, cubic, unilamellar, or bilamellar macromolecular structures. Furthermore, differential scanning calorimetry (DSC) is a thermal analysis tool used to identify phase transitions [4] for LNPs. Both the SAXS and cryo-tem are powerful tools for characterization of LNP structure as a function of formulation manufacture.

In combination of our analytical tools and formulation composition we can generate LNPs of distinct lipid phases. A shift from lamellar to hexagonal shows to correspond to an improvement in potency. A difference in compositional heterogeneity in LNPs is observed to have significant difference in bioperformance. These overall changes in formulation or process are required frequently throughout preclinical, clinical and even sometime post market approval. Henceforth, understanding the key product attributes to in-vivo performance is essential. The analytical tools guide the formulation composition selection, manufacturing process development and ensure manufacturing process consistency.

1. Roces et al., Pharmeceutics, 2020, 12, 1095
2. Gindy et al., Expert Opin. 2012, 9, 171
3. Gindy et al., Langmuir 2014, 30, 16, 4613
4. Eygeris et al., Nanoletters, pending

630409 Dynamic Optimization of an Ultrafiltration System for the Concentration of Monoclonal Antibody Solutions Under Uncertainty

Abstract:

Monoclonal antibodies (mAbs) are a very important type of therapeutic agent with myriad applications in several therapeutic areas, such as immunology, oncology and infectious disease. These immunoglobins owe their success to their higher efficacy, specificity and safety compared to conventional drugs, which is a direct consequence of their mechanism of action. Unfortunately, all these benefits come at a price, namely, mAbs are much more expensive than conventional drugs. In fact, they may cost between 1,000 and 50,000 $/g.

One of the reasons why monoclonal antibodies are extremely expensive is their complex and low-yield production process, which includes several challenging manufacturing steps, namely, an initial fed-batch fermentation followed by centrifugation, depth filtration, several chromatographic purifications, virus filtration and, finally, concentration by ultrafiltration. Therefore, one of the most direct ways to reduce the average cost of monoclonal antibodies is to optimize their production process, in terms of both processing sequence and operating conditions at which the individual manufacturing steps are carried out.

Several authors have proposed optimal design configurations for mAb production processes (Liu et al., 2016) and computed the optimal operating conditions at which the fermentation and chromatographic separations should be carried out (Alhuthali et al., 2019; Papathanasiou et al., 2017). Some of these contributions have also analyzed the impact of model uncertainty on the results of their design and on the associated dynamic optimization calculations (Liu and Gunawan, 2017). Surprisingly, the ultrafiltration step has received very little attention so far, even though the typical yield of this operation may be as low as 90 % (as the monoclonal antibody is being concentrated, a significant amount of it forms a hydrogel that sticks onto the ultrafiltration membrane). Therefore, this contribution investigates the extent to which the systematic use of stochastic dynamic optimization techniques can help increase the efficacy/yield of the ultrafiltration step and reduce its duration.

More specifically, we first develop a hybrid mechanistic-statistical model of the ultrafiltration device, which is comprised of a simplified (yet first-principles inspired) model of the physical system, augmented with a Bayesian neural network (this type of model retains all the benefits of first-principles models, learns from experimental/process data as data-driven models, and provides point estimates of the process variables of interest as well as estimates of their degree of uncertainty). Then, we train this hybrid model on experimental data, collected at several different operating conditions, using advanced optimization-driven uncertainty quantification algorithms. Finally, we make use of the trained hybrid model and of a stochastic dynamic optimization algorithm to optimize the operating conditions of the ultrafiltration system under uncertainty with two principal goals, namely, maximization of the expected process efficacy/yield, measured as final mAb concentration/recovery, and minimization of the ultrafiltration time.

The results of our calculations are extremely promising, in that we can both improve the final mAb concentration/recovery and reduce the ultrafiltration time without violating any operational constraints, even in the presence of significant model uncertainty.

References

Alhuthali, S., Fadda, S., & Kontoravdi, C. (2019). Constrained Optimisation of Cell Culture Feeding Strategy and Temperature Shift Duration to Enhance Monoclonal Antibody Titre and Purity. 2019 AIChE Annual Meeting, Orlando (FL).

Liu, Y., & Gunawan, R. (2017). Bioprocess optimization under uncertainty using ensemble modeling. Journal of biotechnology, 244, 34-44.

Liu, S., Farid, S. S., & Papageorgiou, L. G. (2016). Integrated optimization of upstream and downstream processing in biopharmaceutical manufacturing under uncertainty: a chance constrained programming approach. Industrial & Engineering Chemistry Research, 55(16), 4599-4612.

Papathanasiou, M. M., Quiroga-Campano, A. L., Steinebach, F., Elviro, M., Mantalaris, A., & Pistikopoulos, E. N. (2017). Advanced model-based control strategies for the intensification of upstream and downstream processing in mAb production. Biotechnology progress, 33(4), 966-988.

629569 Confocal Imaging of Protein Gel Layer Formation during Tangential Flow Filtration to Inform Process Conditions Reducing Protein Losses and Increasing Protein Concentration

Abstract:

There are about 570 antibody therapeutics at various phases of clinical trials across the biopharmaceutical industry with 51 mAbs and 11 biosimilars in late-stage trials (Kaplan and Reichart, 2019). Of these 93% are produced in the US and Europe, and half are based on fully human genetic sequences (Grilo and Mantalaris, 2019). Therapeutic levels of monoclonal antibodies (mAbs) lie between 5 to 1000 mg per patient per treatment (Garidel et al., 2017; Maruthamuthu, 2020). Hence, mAbs must be concentrated to high levels (generally in the range of 150 to 200 mg/mL) to achieve therapeutic concentrations and/or reduce dosage volume (Garidel et al., 2017).

Tangential flow filtration (TFF) is the major unit operation in biopharma manufacturing for concentration of proteins. TFF uses shear forces along the membrane surface to limit protein accumulation on the membrane during the ultrafiltration process that occurs at pressures of up to 60 psig. Decreases in flux through the membrane correlate to increases in viscosity and osmotic pressure that occur when bulk protein concentration increases (Baek and Zydney, 2018). In addition, protein accumulation on the membrane's surface results in a gel layer that causes resistance to flux. Mechanisms of gel layer formation are complex and may depend on flow patterns, cross-membrane velocities, and biophysical properties that are a function of protein concentration, molecular weight, and pI.

We report a prototype confocal membrane chamber that enables observation of protein layer development on the surface of the membrane and along the length of the ultrafiltration module. This optical flow cassette was fabricated from a commercially available TFF cassette (Ultracel 30 kD_a pellicon 3) to expose the feed screen and the membrane underneath. A thin, optically transparent plexiglass cover placed over the surface of the feed screen forms a viewable feed channel enabling confocal imaging within a 4mm, optical path length. Flow passes across the membrane through a 1 mm high channel (distance from window to membrane surface). The cover was glued in place along the edges to form a pressure tight seal resulting in an operational replica of a bench-top TFF cassette with a membrane surface area of 44 cm^2.

A solution of Bovine IgG (50 mg/mL; in 10 mM citrate buffer with 50 mM NaCl, pH 5.5) was doped with a fluorescently labeled protein and concentrated in this prototype optical flow cell. This protein was used since its molecular weight of 150 kD is in the same range of molecular weights of many therapeutic mAbs presented in the literature, with a pI of 7.2 that lies in the mid-range of therapeutic mAbs whose pI's range from 6 to 9. The accumulation of protein on the membrane surface was captured with z-stack imaging. Together with separate measurements of flux, retentate concentration, transmembrane pressure, and flow rates, these images are helping to inform development of a computational model for predicting conditions that maximize protein concentration and recovery (Rossi et al., 2021). In this paper, we report visualization and characteristics of a gel layer at TFF membrane surfaces for bulk protein concentrations ranging from 50 to 200 mg/mL for process conditions using confocal microscopy.

References

Baek, Y. and A. Zydney. (2018). Intermolecular interactions in highly concentrated formulations of recombinant therapeutic proteins, Current Opinion in Biotechnology, *53*, 59-64.

Garidel, P., Kuhn, A. B., Schäfer, L. V., A. R., Karow-Zwick, & M. Blech. (2017). High-concentration protein formulations: how high is high? European Journal of Pharmaceutics and Biopharmaceutics, 119, 353-360.

Grilo, A. L., A. Mantalaris. (2019). The increasingly human and profitable monoclonal antibody market, *Trends Biotechnol.*, *37*, 9-15.

Kaplon, H., J. M. Reichart. (2019). Antibodies to watch in 2019, Mabs, *11*, 219-238.

Maruthamutu, M. K., S. R. Rudge, A. Ardekani, M. R. Ladisch, and M. S. Verma. (2020). Process Analytical Technologies and Data Analytics for the Manufacture of Monoclonal Antibodies, Trends in Biotechnology, *38*(10), 1169-1186.

Rossi, F., J. Zuponcic, E. Ximenes, S. Geng, Y. Taa, V. Corvari, M. Ladisch, R. Reklaitis. (2021). "Dynamic Optimization of an Ultrafiltration System for the Concentration of Monoclonal Antibody Solutions under Uncertainty," Abstract submitted to Annual AIChE meeting.

629177 End-to-End Reconfigurable Process Development for the Cancer Drug Lomustine

Abstract:

Anti-cancer therapeutics represent a specialized portion of the pharmaceutical market. The recipients of these medications are often facing substantial health challenges, but supply chain issues may negatively impact patient access. Our research team at Purdue University a small-scale integrated, modular, reconfigurable platform was designed and constructed as a proof-of-concept platform that can can be used to meet the market demand for Lomustine, an anti-cancer drug first produced in 1976 that is a first line treatment for glioblastoma and a second line agent for Hodgkin's lymphoma[1]. In 2014, a change in manufacturing rights caused the Lomustine price to increase from $50 to $768 per capsule[2] in subsequent years, which helped to garner attention from the worldwide community. This work details the successful integration of a small-scale end-to-end process for continuous and hybrid (combined continuous-batch) manufacturing process alternatives for Lomustine.

In this work, enhanced Quality by Design (QbD) techniques – including multivariate experiments and process analytical technology (PAT) tools – were used to determine the impact of various process parameters on design space and drug substance critical quality attributes (CQAs)[3]. Additionally, we eliminated a reagent used in the novel flow synthesis pathway for Lomustine to improve process safety,[4] and carried out the synthesis in low-cost, tubular fluorinated propylene ethylene (FEP) reactors. The work also highlights our upscaled Lomustine synthesis and process intensification with an integrated continuous solvent switch distillation (CSSD) module and flow crystallization module. After discovering that crystallization was not achieved directly from the reaction mixture, we designed the CSSD to operate under vacuum, enabling the continuous evaporation of the reaction solvent, while continuously feeding in the crystallization solvent. Nitrogen-segmented flow crystallization was implemented in a low-cost PTFE, tubular crystallizer to provide for final drug substance isolation. The fully continuous process alternative was of particular interest because it provides enhanced manufacturing safety of this DNA alkylating agent by reducing the handling of Lomustine, as well as its intermediate and associated reagents, prior to the crystallization step. Experimental validation of other selected process alternatives demonstrate the ability of PharmaPy to capture platform dynamics[5] and the reconfigurability of the manufacturing platform. Both batch and continuous unit operation modes were considered to find an "end-to-end optimal" configuration that does not require fully continuous operation.

Additionally, integration of PAT for process monitoring is incorporated within this study and provides key insight into this rarely studied, Lomustine manufacturing process. Reaction monitoring with Mettler Toledo's React IR and automated sampling through EasySampler allowed for reaction component monitoring with compounds that are active at the same UV wavelength range. Batch crystallization monitoring was achieved with the use of various PAT tools including: focused beam reflectance measurement (FBRM), UV-Vis spectroscopy, and particle vision and measurement (PVM). The use of various PAT tools helped to define the operating space and determine key parameters for model development[6].

In summary, this presentation depicts the experimental validation of various manufacturing process alternatives for the anti-cancer drug Lomustine. This work showcases novel approaches in engineering to manufacture a high value compound with low-cost process materials. Enhanced QbD approaches were used to determine operating conditions that result in desired CQAs for isolated Lomustine product. Finally, manufacturing process design and intensification with small-scale modules integrated with PAT tools provided valuable insight into process dynamics and find an "end-to-end optimal" configuration.

References

1. Lee, F. Y.; Workman, P.; Roberts, J. T.; Bleehen, N. M., Clinical pharmacokinetics of oral CCNU (lomustine). *Cancer Chemother. and Pharmacol.* 1985, *14* (2), 125-131.
2. Loftus, P. Cancer Drug Price Rises 1,400% With No Generic to Challenge It *The Wall Street Journal* [Online], 2017. https://www.wsj.com/articles/cancer-drug-price-rises-1400-with-no-generic-to-challenge-it-1514203201 (accessed Mar 15, 2019).
3. U.S. Food and Drug Administration, Guidance for Industry Q8(R2) Pharmaceutical Development. 2009. https://www.fda.gov/regulatory-information/search-fda-guidance-documents/q8r2-pharmaceutical-development (accessed Mar 29, 2021).
4. Jaman, Z.; Sobreira, T. J. P.; Mufti, A.; Ferreira, C. R.; Cooks, R. G.; Thompson, D. H., Rapid On-Demand Synthesis of Lomustine under Continuous Flow Conditions. *Org. Process Res. & Dev.* 2019, *23* (3), 334-341.
5. Casas-Orozco, D.; Laky, D.; Hur, I.; Mackey, J.; Mufti, A.; Wang, V.; Abdi, M.; Feng, X.; Wood, E.; Reklaitis, G. V.; Nagy, Z. In *Use of PharmaPy for digital twin construction of a process for the manufacturing of Lomustine*, AICHE, 2021.
6. Casas-Orozco, D.; Laky, D.; Mackey, J.; Mufti, A.; Reklaitis, G. V.; Nagy, Z. In *Chemometric techniques for the combined calibration/parameter estimation of pharmaceutical drug substance manufacture*, AICHE Annual Meeting, 2021.

623572 Continuous Recycling of an Immobilized Biocatalyst in a Solids Forming Reaction By Magnetic Separation

Abstract:

Enzymatic reactive crystallization is a promising green synthesis route for the production of semi-synthetic β-lactam antibiotics. End-to-end continuous manufacturing is one method to assist in careful control of the stable operating point that reactive crystallization requires. For β-lactam antibiotic synthesis, simulations have shown that use of continuous reactive crystallization can be optimized to increase productivity and conversion compared to batch synthesis [1]. One major limitation is in the use of a solid catalyst (immobilized enzyme) in the presence of crystallizing API and the resulting solid-solid separation that is required to purify the API and recycle the catalyst.

Solid-solid-liquid separations are difficult and uncommon design problems that are typical to the mineral industry, but rarely used in the pharmaceutical industry. One possible solution is the use of magnetic separation to separate enzyme immobilized to paramagnetic material from the nonmagnetic crystalline product. While magnetic separation has been implemented in several prior systems, including with the enzyme penicillin G acylase (PGA) as used in this study [2-4], the application detailed in this work is unique due to the continuous nature of the separation and its use at the pharmaceutical pilot plant scale.

In this study, a continuous magnetic separator was designed and optimized using iterative prototyping. Processing conditions that impact separation efficiency and solids throughput were evaluated using a Buckingham Pi analysis. Further optimization of the device was performed by separating the magnetic enzyme immobilization platform from nonmagnetic beads under flow rates from 3 – 15 mL/min. Separation efficiencies of >99% were obtained at flow rates and solid densities comparable to the current pilot plant used for cephalexin synthesis. The separator was then demonstrated in a batch operation for the separation of carrier and cephalexin crystals under conditions that mimicked the expected output from the pilot plant. Next, operation for the separator was demonstrated continuously for one and five hours, again with simulated pilot plant output conditions. Finally, the device was demonstrated continuously for 18 hours to separate and recycle immobilized PGA from cephalexin crystals generated from the enzymatic reactive crystallization of cephalexin at the pilot plant scale.

[1] McDonald, M.A., et al., *Continuous reactive crystallization of β-lactam antibiotics catalyzed by penicillin G acylase. Part I: Model development.* Computers & Chemical Engineering, 2019. **123**: p. 331-343

[2] Luo, X., and Zhang, L. (2010) Immobilization of penicillin G acylase in epoxy-activated magnetic cellulose microspheres for improvement of biocatalytic stability and activities, *Biomacromolecules 11*, 2896-2903.

[3] Netto, C. G., Toma, H. E., and Andrade, L. H. (2013) Superparamagnetic nanoparticles as versatile carriers and supporting materials for enzymes, *Journal of Molecular Catalysis B: Enzymatic 85*, 71-92.

[4] Liu, R., Huang, W., Pan, S., Li, Y., Yu, L., and He, D. (2020) Covalent immobilization and characterization of penicillin G acylase on magnetic Fe_2O_3/Fe_3O_4 heterostructure nanoparticles

prepared via a novel solution combustion and gel calcination process, *International Journal of Biological Macromolecules*.

622860 Spray Drying Proteins: Advances in Process Understanding Using the Model Protein Human Serum Albumin

Abstract:

Biopharmaceuticals, such as monoclonal antibodies and protein therapeutics, steadily increase in importance in the pharmaceutical industry. Advances in research, with the need for biopharmaceutical products to treat various diseases speed up the development of new protein-based drugs and the production of these valuable entities. Transferring proteins from liquid state to solid state has already been shown to be beneficial for the proteins' stability and enables milder storage condition [1]. So far, freeze-drying is the gold standard when it comes to transferring protein solutions to protein powder on an industrial scale. However, freeze drying is a cost-and, energy-intensive process with common drying times from hours to days [2,3]. Spray drying, as an alternative, enables continuous protein drying with flexible process design, at atmospheric conditions and milder temperatures enable avoidance of temperature induced denaturation. However, commonly spray dried powders exhibit larger residual moisture as drying takes place in the order of seconds [2,3].

Therefore, we investigated at first optimal process parameter sets for a spray drying process at mild temperatures. With the findings, different spray drying process parameter sets for the model protein human serum albumin (HSA) were selected for optimal spray drying at low residual moisture and high yields. Spray drying HSA with these sets was performed under low bioburden conditions with the drying gas being nitrogen. Further, the drying process was monitored in terms of inlet drying gas, and outlet drying gas conditions, as well as temperatures of the feed and product. Besides the process investigation, the impact of the process sets on the properties of the protein is tested with a standardized set of methods after spray drying in solid and resolubilized state. Further, we set up a model to gain further insight and enable the linkage of process knowledge and protein properties for future process development.

The talk will highlight the lessons learned from the primary process parameter investigation at mild conditions and from spray drying the model protein HSA. Process knowledge will be included gained from investigations on the influence on the protein properties. A comparison will be drawn with spray dried HSA and the developed model used to depict the spray drying process. We aim to support development for spray drying as alternative technology and industrialization to fully exploit its benefits for protein-based therapeutics.

Literature Cited

1. Maltesen MJ, van de Weert M. Drying methods for protein pharmaceuticals. *Drug Discov. Today Technol.* 2008;5(2-3):e81-8.
2. Ratti C. Freeze-Drying Process Design. In: Ahmed J, Rahman MS, editors. *Handbook of Food Process Design*. Hoboken: Wiley-Blackwell [Imprint]; John Wiley & Sons, Incorporated, 2012:621–647.
3. Langford A, Bhatnagar B, Walters R, Tchessalov S, Ohtake S. Drying of Biopharmaceuticals: Recent Developments, New Technologies and Future Direction. *Jpn. J. Food Eng.* 2018;19(1):15–24.

624667 Novel Vaginal Pessary Platform Via 3D-Printing: Concomitant Mechanical Support and Drug Delivery

Abstract:

Currently, pessaries are removeable, drug-free polymeric structures that are inserted into the vagina to support the pelvic organs, including the urethra, the bladder and the uterus. They are used to treat urinary incontinence, especially stress urinary incontinence, which affects more than 40% of women worldwide (1). Although pessaries offer great advantages, such as easy self-administration and immediate relief of symptoms (namely involuntary urine loss) (2), they are not associated with great patient adherence (3). This is mainly due to the fact that commercial pessaries are made of silicon via injection molding and are thus, available only in a few sizes and shapes. We just recently showed that 3D-printing, more specifically material extrusion-based additive manufacturing, is an emerging manufacturing technique that circumvents these hurdles and allows for pessary personalization (4). In a subsequent process step, selected 3D-printed pessaries were loaded with drug via immersion (5). In an attempt to avoid issues associated with immersion processes, including non-uniform drug distribution, non-reproducible drug content and the application of organic solvents, we report the direct manufacturing of personalized, drug-loaded pessaries – particularly, incontinence rings – via a combination of filament extrusion and 3D-printing for the first time. Thereby, not only mechanical support but also a concomitant pharmacological therapy is provided, yielding a new generation of pessaries. Thermoplastic polyurethanes (TPUs) of different grades (varying hardness and hydrophobicity) served as polymeric carriers. In a first step, drug-free incontinence rings were prepared and tested for their mechanical characteristics. Both, the maximum compression force at a given deformation – which is a measure for the extent of mechanical support the incontinence ring provides in the vaginal cavity – and the component compression stiffness - indicating how easily the incontinence ring can be compressed for manual self-insertion into the vagina - were a function of i) the TPU grade, ii) the cross-sectional diameter and iii) the infill. For a set cross-sectional diameter (i.e., 7 mm), infills (i.e., 10-100%) were identified for each TPU grade to achieve favorable mechanical characteristics. In a second step, these parameters were applied to 3D-print incontinence rings comprising 10 wt.% of the model drug acyclovir (ACV). After 3D-printing, the ACV solid state was affected by the TPU grade (due to different hydrophobicities) and/or the nozzle temperature. Clearly, the in-vitro drug release properties were influenced by the drug solid state. However, the key factors to trigger the in-vitro ACV release from the 3D-printed incontinence rings – including the amount of ACV released per time unit, the duration of ACV release, and the release mechanism – were the interaction of the TPU polymer with the dissolution medium (swelling capacity) and the infill (interior porosity). Summarizing, our results demonstrate that TPU is a suitable polymer to build – together with 3D-printing – the platform for novel drug-loaded pessaries. The critical quality attributes of the novel pessary product, namely the mechanical properties and the in-vitro drug release characteristics are, however, not limited to the polymer material properties. Instead, the 3D-printing process allows for their adjustment within a broad range, which is the basis for personalized drug products.

References

1. Reynolds WS, Dmochowski RR, Penson DF. Epidemiology of stress urinary incontinence in women. Curr Urol Rep. 2011;12(5):370–6.
2. Al-Shaikh G, Syed S, Osman S, Bogis A, Al-Badr A. Pessary use in stress urinary incontinence: a review of advantages, complications, patient satisfaction, and quality of life. Int J Womens

Health. 2018;10:195–201.
3. Jones KA, Harmanli O. Pessary Use in Pelvic Organ Prolapse and Urinary Incontinence. Rev Obstet Gynecol. 2010;3(1):3–9.
4. Spörk M, Arbeiter F, Koutsamanis I, Cajner H, Katschnig M, Eder S. Personalised urethra pessaries prepared by material extrusion-based additive manufacturing. *In preparation.*
5. Koutsamanis I, Paudel A, Alva Zúñiga CP, Wiltschko L, Spoerk M. Novel 3D-printed thermoplastic copolyesters for implantable drug delivery applications. *In preparation.*

624055 Detecting and Evaluating Irregular Objects in OCT Images with Unsupervised Machine Learning

Abstract:

Recently, optical coherence tomography (OCT) has gained significant attention in the pharmaceutical industry due to its capability for real-time monitoring of coating processes, such as film coating of tablets and pellets. Accurately detecting the boundaries of objects in OCT is crucial to a meaningful evaluation of the image. Thus far, the focus has been on evaluating the mean coating layer thickness, using algorithms which presume a spherical, smooth shape of the product or - in the case of Convolutional Neural Networks (CNNs) - require large amounts of hand-annotated training data specific for each product. These methods reach their limits, if the shapes depicted on the OCT images are not spherical or cannot be trained easily, as is the case of irregularly shaped tables or pellets.

Significant work has been undertaken to address these challenges and a method was developed, which is able to evaluate a variety of product shapes by adjusting parameters accordingly, hence avoiding lengthy manual annotation of many OCT images. This new algorithm makes use of unsupervised machine learning and applies it to OCT image segmentation. Specifically, by working with an iterative spatial clustering approach, we are capable of capturing details, such as speckles and artifacts, while discarding noise. This technological approach enables us to accurately evaluate the surface roughness of non-smooth, non-spherical products. Furthermore, by distinguishing tablet features from image noise we facilitate an accurate representation of a coating layer or a product's homogeneity. Ongoing research aims to utilise the results of these new algorithms by investigating a broader range of pharmaceutical products, beyond tablets and pellets.

626155 Direct Methylation of Cobaloxime By Low Temperature Methane Plasma in a Multiphase DBD μ-Plasmatron

Abstract:

Plasma-assisted heterogeneous catalysis and its potential in synthetic chemistry is being studied for the highly active species generated at mild conditions and high energy efficiency compared to traditional synthetic methods, as well as being environmentally green owing to the elimination of toxic solvents during synthesis processes. [1–5] Micro dielectric-barrier-discharge (DBD) plasma reactors are suitable devices for the study of plasma synthetic chemistry since they offer homogeneous non-thermal plasma along with reduced transport limitations between plasma and liquid phase. [6,7] Cobaloximes are receiving increasing attention for their roles in photoredox/organometallic cooperative catalysis.[8,9]

A cobaloxime molecule is chosen as an example organometallic molecule for our study. Experiments were conducted in a multiphase μ-plasmatron, in which the multiphase mixing was previously evaluated. [10] Generation of methyl radicals was confirmed by in-situ optical emission in addition to circuit current measurement, which facilitated the study of plasma catalytic reactions. Directly methylation of a cobalt compound by methyl radicals was confirmed by its characteristic Raman shifts, which is exciting for it explored the potential of using methane as a feedstock in organometallic synthesis. The short reaction residence time also demonstrated "flash chemistry" made possible in a multiphase μ-plasmatron. The above work will be presented and discussed.

1. Bogaerts. *J. Phys. D. Appl. Phys.* **53**, 443001 (2020).
2. Mariotti. *J. Phys. D. Appl. Phys.* **43**, 323001 (2010).
3. Anastasopoulou. *J. Phys. D. Appl. Phys.* **53**, 234001 (2020).
4. Mishin. *Russ. J. Gen. Chem.* **85**, 1209–1221 (2015).
5. Nozaki. *Catal. Today* **211**, 29–38 (2013).
6. Losey. *Ind. Eng. Chem. Res.* **40**, 2555–2562 (2001).
7. Wengler. *React. Chem. Eng.* **3**, 930–941 (2018).
8. Sun. *Nat. Chem.* **10**, 1229–1233 (2018).
9. Johnston. *Nature* **536**, 322–325 (2016).
10. Liu. *J. Phys. D. Appl. Phys.* **54**, 194003 (2021).

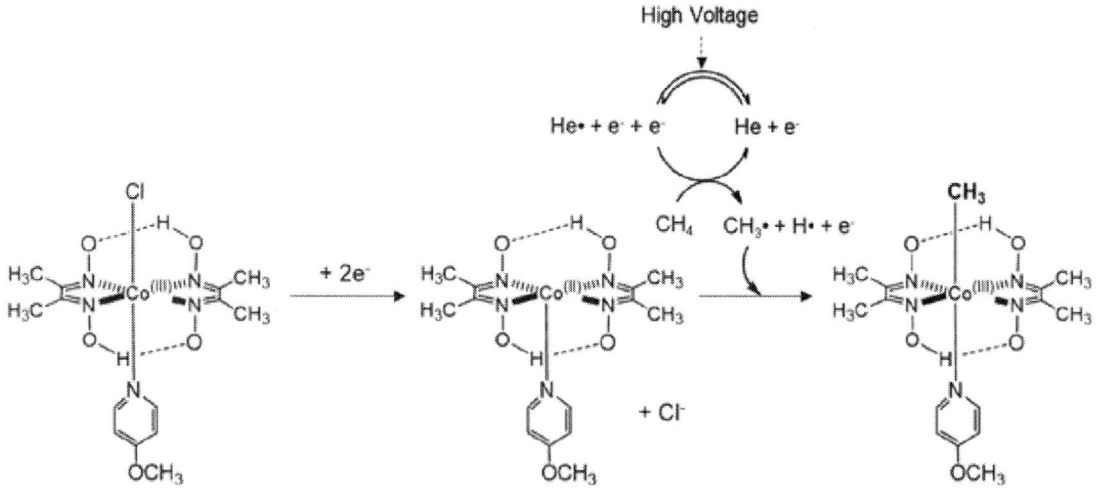
High Voltage
He• + e⁻ + e⁻
He + e⁻
CH4
CH3• + H• + e⁻
+ 2e⁻
+ Cl⁻
OCH3

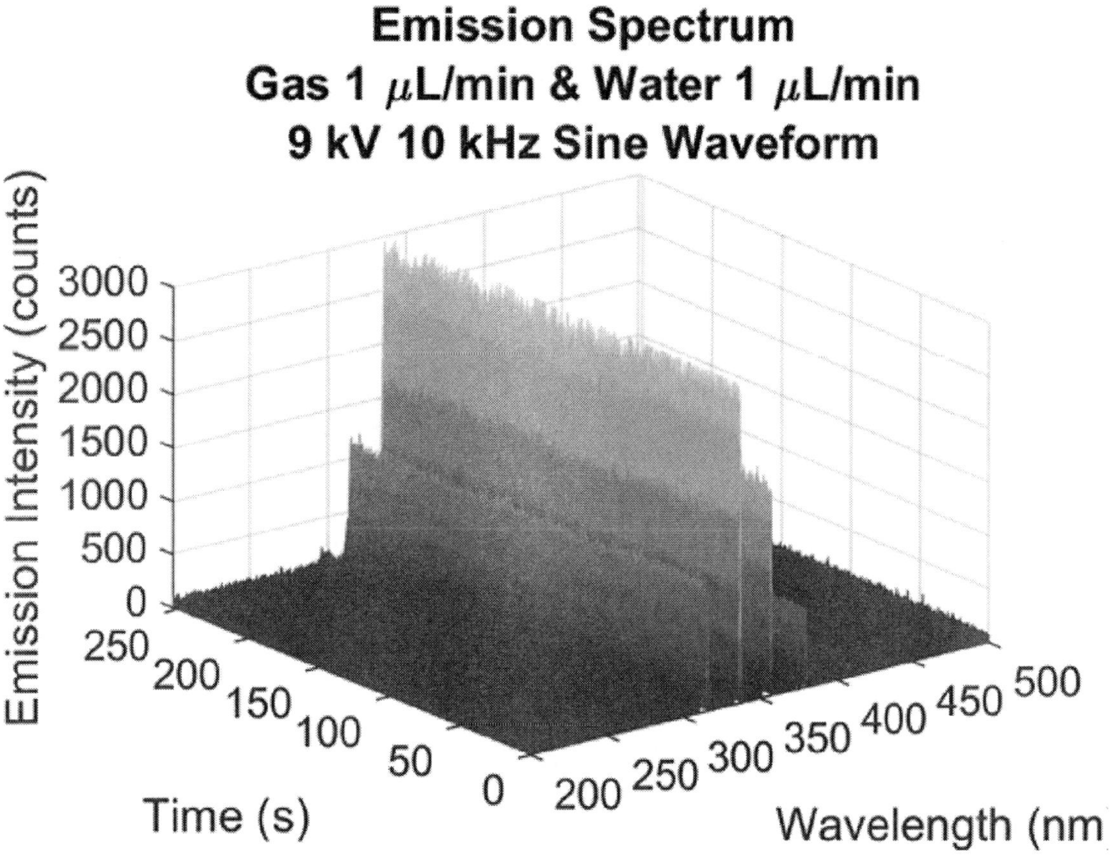
Emission Spectrum
Gas 1 μL/min & Water 1 μL/min
9 kV 10 kHz Sine Waveform
Emission Intensity (counts)
3000
2500
2000
1500
1000
500
0
250
200
150
100
50
0
Time (s)
200
250
300
350
400
450
500
Wavelength (nm

629622 Co-Hydrothermal Carbonization (Co-HTC) of Biomass-Plastic Blend: Enhancement of Fuel Quality

Abstract:

Disposal of waste plastics and biomass have been alarming issues for the U.S. over the past few years. A sustainable circular bio-economy has been a prime need to shift towards alternative fuels from traditional fossil fuels. Unfortunately, waste biomass contains high inorganics and low carbon content that makes biomass unfavorable for co-firing. On the other hand, waste plastics e.g., discarded mattress has low bulk density and contains high nitrogen content. Both feedstocks need preprocessing prior to use as fuel. This study focused on converting waste corn-stover into carbon-rich value-added biofuel through blending with waste polyurethane. The biomass-plastic blend was mixed at 95%:5%, 90%:10%, and 85%:15% (w/w dry basis) ratios, and co-hydrothermally carbonized (co-HTC) at 200, 230, and 260 °C. The ultimate analysis was done to determine the elemental composition in the resulting hydrochars and compared with the control runs (HTC of biomass and plastics at the corresponding temperatures). The thermogravimetric analysis was performed both in nitrogen and oxygen environments to observe the thermograms and combustion properties of the hydrochars. The higher heating value was obtained to observe the upgrade in the fuel value via Co-HTC. It is found that the total elemental carbon content of co-HTC hydrochars are higher than the HTC of biomass-only. On the contrary, the total elemental oxygen and nitrogen content also drop in polyurethane which augment the overall fuel value of the blended-biomass. Additionally, the ash content significantly decreases in both biomass and polyurethane. It was found that the fuel value of co-HTC hydrochar increases up to 26 MJ/kg, which is like lignite coal. Low sulfur, nitrogen, and oxygen content of the co-HTC hydrochar along with high energy content make co-HTC hydrochar a viable fuel. This study shows a sustainable route to produce blended biomass-plastic hydrochars with enriched fuel value from waste agricultural residue and plastics.

622229 Solvothermal Liquefaction of Mixed Waste Plastics: Effect of Solvents at Sub-and Supercritical Conditions

Abstract:

Plastics are being used widely in different applications these days due to their advantages such as corrosion resistance and also being light yet stable and having excellent insulating properties. These benefits have made plastics among the leading materials in different usage models. Despite all the aforementioned benefits, plastics are also blamed for a series of health and safety problems in addition to widespread pollution of the planet. Waste to liquid is one of the main solutions identified to overcome these devastating problems. This study aims to investigate the effects of different solvents such as ethanol, water, n-hexane, and toluene in subcritical and supercritical conditions to find the influence of polarity of solvents on depolymerization of plastics. Additionally, the effects of reaction parameters such as temperature (200, 275, and 350 °C) and time (60 -180 min) on the solvothermal liquefaction (STL) of mixed plastics combined from (polypropylene, low-density PE, polystyrene, and polyurethane) on the oil yield, composition, and quality of the bio-oil formed from STL are evaluated through ultimate analysis, proximate analysis, chemical composition, gas chromatography mass spectroscopy, proton nuclear magnetic resonance, thermogravimetric analysis, and Fourier-transform infrared spectroscopy of bio-oil. The reaction temperature was found to have a positive effect on biooil yield for the solvents; however, it is more significant for n-Hexane solvent in supercritical condition. This study was a step-forward towards investigating the effects of such solvents in order to figure out the influence of polarity of solvents on physicochemical properties and oil yield from mixed plastics.

628982 Selective Ring Opening of Decalin over Ion-Exchanged Ir Zeolites

Abstract:

Thermal DeOxygenation (TDO) is a thermal process that converts organic acids from cellulose hydrolyzation into a low-oxygen bio-oil containing substituted naphthalene compounds. The high aromatic content is desirable in gasoline fractions, but middle-distillate cuts, such as jet fuel and diesel, require upgrading via hydrogenation and ring opening to achieve better combustion characteristics. Previous research has demonstrated that TDO oil could be hydrogenated over high-surface-area nickel catalysts to increase the combustion characteristics, but the cetane number was still below specifications.

This research focuses on the ring opening of decalin, a bicyclic molecule with a chemical structure similar to many of the compounds found in hydrogenated TDO oil. Because there is no commercial process for these reactions, there has been little catalyst development in ring opening (hydrodecyclization), especially in the production of paraffins/isoparaffins from cycloparaffins. In our down-flow trickle bed reactor, we use various catalysts to ring open decalin under a variety of operating conditions. In this research, we analyze the effect of catalyst acidity, ion exchanges, and decalin stereochemistry on the reaction products.

Our research has shown that the concentration of Brønsted acid (BA) sites is crucial to determining the reaction pathway (see image). Catalysts with high quantities of BA sites isomerize decalin into unwanted skeletal isomerization products (SkIPs), whereas low concentrations of BA sites push the reaction towards ring opening products (ROPs). The catalyst BA was tuned by ion-exchanging various alkali metals into the zeolite at various concentrations and temperatures to optimize desired products. Our Ir/Cs-exchanged β-zeolites have been shown to produce products with ROP selectivities in excess of 80%, while limiting SkIP formation to less than 5%.

Our research also showed that the decalin stereochemistry played a crucial role in determining the reaction mechanism. *Trans*-decalin was more likely to produce branched SkIPs, whereas *cis*-decalin preferentially produced ROPs.

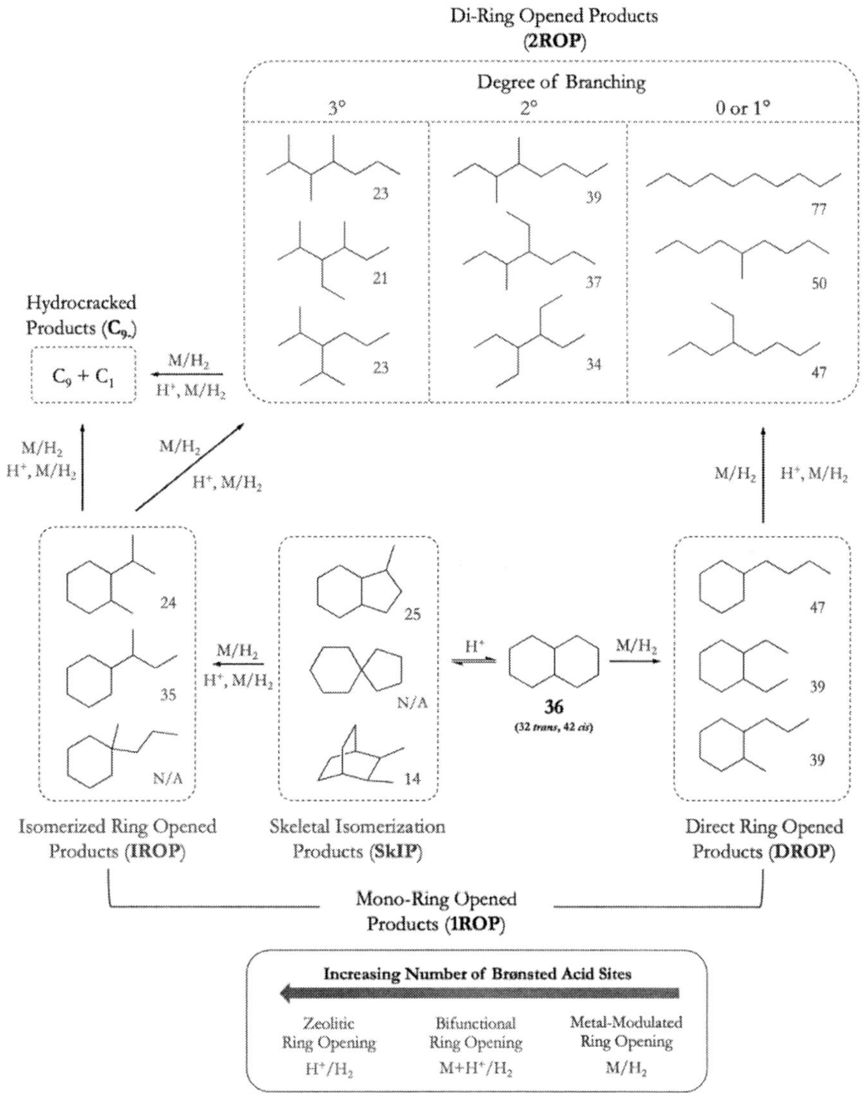

Ring opening and isomerization products formed from decalin over various catalysts. Cetane numbers of chemicals are adjacent to their structure.

630830 Forced Thermal Oscillations in Catalytic Microreactors

Abstract:

Engineers and scientists have spent decades understanding catalysts and have generally come to the conclusion that kinetic rate optimization is fettered by the thermodynamics of the surface-species interactions. This limit is established by Sabatier's rule and dictated by the scaling relations. While for years this limitation was taken at face value and deemed insurmountable, recent studies have begun finding creative ways to better optimize or even overcome this thermodynamic limitation. Reaction dynamics, or operating a reactor in a controlled, oscillatory fashion, is one such way that researchers have begun to overcome these limitations. In this work I developed a microreactor system capable of oscillating the reaction temperature at frequencies up to 10Hz through rapid resistive heating to explore the effects of dynamic systems.

632048 Nickel-Iron Catalysts for Low Temperature Dry Reforming of Methane

Abstract:

Dry reforming of methane (DRM) converts two important greenhouse gases, methane and carbon dioxide, in a single reaction to produce syngas. Nickel-based catalysts have been extensively studied for DRM. However, monometallic Ni catalysts deactivate because of coking. We are motivated to include earth-abundant promoter metals to inhibit coke formation and studied series of bimetallic nickel-iron catalysts supported over P-25 TiO_2 at 550°C and atmospheric pressure. Ni-Fe catalysts were prepared by wet impregnation and co-precipitation methods. The total metal loading was maintained at 10 wt% with different ratios of Ni/Fe. DRM activity results showed that co-precipitation Ni-Fe/TiO_2 catalyst with Ni/Fe ratio of 3:1 is optimum in terms of CH_4 and CO_2 consumption showing similar reaction rates to Ni/TiO_2 and minimal coke deposits simultaneously. H_2-TPR results showed higher reducibility and enhanced metal and support interaction in Ni-Fe/TiO_2 synthesized by co-precipitation catalysts than that of impregnation catalysts. The activity of CH_4 and CO_2 was higher over Ni-Fe/TiO_2 co-precipitation catalysts than impregnation catalysts. XPS analysis of reduced catalysts showed formation of interacting bimetallic Ni-Fe species over co-precipitation catalysts, while no such Ni-Fe interactive species formed over impregnation catalysts. Spent catalysts which are characterized by TGA-DTG and Raman spectroscopy, showed decrease in coke deposits by 20 times over Ni_3Fe_1/TiO_2 in comparison to monometallic Ni/TiO_2 catalysts. *In-situ* DRIFTS analysis revealed that DRM reaction intermediate species including hydroxyls (OH^*) and carbonates (COO^*) oxidized coke precursors over Ni-Fe/TiO_2 catalysts.

634114 Keynote Talk: Online Measurements for the Petrochemical Industry: Industry 4.0 Trends and Unmet Needs

Abstract:

The primary objectives of the petrochemical manufacturing industry are to operate plants safely, profitably, and reliably. To achieve these objectives, several constraints need to be managed – such as raw material yield, manufacturing costs, and utility requirements. Online measurements via instruments that measure the physical properties and analyzers that measure the chemical properties of the process are critical to the optimized management of these constraints.

Globally, Dow owns and supports > 1 million instruments and > 20,000 online analyzers. Data from these measurements are indispensable to the operation and control of our plants. In this talk we will discuss some of our key requirements for both classes of online measurements, share examples of meaningful advances afforded by Industry 4.0, and articulate key challenges that remain – a call for continued innovation.

628357 Woodchip Moisture Content Estimation Using Short-Range Iot Wi-Fi for the Pulp & Paper Industry

Abstract:

The pulp and paper industry is the third-largest consumer of energy in the US industrial sector thereby leading to tremendous opportunities to improve its energy efficiency and productivity. The pulping process, which converts woodchips into pulp by displacing lignin from cellulose fibers, is one of the most important operations in a pulp and paper mill. For the pulping process, it is important to have real-time knowledge about the moisture content (MC) of the woodchips so that the process can be optimized and/or controlled correspondingly to achieve satisfactory product quality while minimizing the consumption of energy and chemicals. There have been attempts to develop real-time woodchip MC sensors, such as capacitance, resistance, microwave and NIR-based sensors. But these techniques have not been adopted by the industry due to unsatisfactory performance and/or high maintenance requirements that can hardly be met in a manufacturing environment. Also, some of the accurate methods are often lab-based that cannot be implemented online because of time constraint [1]. Due to these limitations, at present, the moisture level is measured only a few times throughout the year in the pulp and paper industry.

To address these limitations, we propose a non-destructive, economic, and robust approach using industrial Internet-of-Things (IIoT) based Wi-Fi and employ channel state information (CSI) to estimate MC in woodchips. We extract CSI by modifying the open-source device drivers for Intel Wi-Fi link 5300 Network Interface Card based on CSITool [2]. CSI contains information about the channel in the form of individual data sub-carriers capturing indoor channel features such as the effect of scattering, fading, and power decay with distance. We hypothesize that when a Wi-Fi signal passes through wood chips, the water or MC in woodchips would cause a corresponding change in the Wi-Fi signal which is then captured in CSI. The goal is to develop a data-driven model to extract the relationship between the change in CSI and the woodchip MC. CSI has been successfully used for moisture classification in wheat [3]. However, compared to wheat moisture classification, woodchip moisture estimation is much more challenging due to the significant heterogeneity in the shape, structure, and arrangement of the woodchips. For example, the woodchip arrangement in the container is expected to have a significant impact on the CSI data. In addition, CSI data have only been used in classifications, which is much easier than regression modeling.

In this work, we propose a regression-based approach to estimate the moisture content in wood chips. To the best of our knowledge, this is the first work to address the continuous estimation of MC using CSI data. Our goal is to develop a robust model that is insensitive to the heterogeneity of the woodchips, particularly the confounding factors such as woodchip shape and the effect of shuffling, to achieve a consistent and accurate MC estimation. While the IIOT devices used are small, low-cost, and rugged to stand for harsh environment, they do have their limitations such as the raw CSI data are very noisy, which can lead to low performing models when directly fed to machine learning algorithms. To address this challenge, our previously developed statistics pattern analysis (SPA) framework [4][5] is used to extract robust and predictive features from raw CSI data. The SPA-based features are then used to develop robust machine learning models for continuous MC estimation in woodchips. In other words, SPA turns messy big data into so-called smart data, which can then be used for accurate sensing. The machine learning techniques studied in this work include artificial neural networks (ANN), support vector machines (SVM), and Gaussian process regression (GPR). We conducted systematic study of these nonlinear regression approaches, and demonstrated that, when combined with SPA, provide a potential solution for real-time woodchip MC estimation for the pulp and paper industry. Specifically, the experimental validation results show highly accurate prediction of more than 20 MC levels in a wide range. In addition to the excellent performance, the proposed approach is also economic and non-destructive, making it a potential technology for industrial implementation in the pulp and paper industry.

[1] R. Govett, A PRACTICAL GUIDE FOR THE DETERMINATION OF MOISTURE CONTENT OF WOODY BIOMASS A Practical Handbook of Basic Information, Definitions, Calculations, Practices and Procedures for Purchasers and Suppliers of Woody Biomass.

[2] D. Halperin, W. Hu, A. Sheth, and D. Wetherall, Tool Release: Gathering 802.11n Traces with Channel State Information, *ACM SIGCOMM Comput. Commun. Rev.*, 2011.

[3] W. Yang, X. Wang, S. Cao, H. Wang, and S. Mao, Multi-class wheat moisture detection with 5GHz Wi-Fi: A deep LSTM approach, in *Proceedings - International Conference on Computer Communications and Networks, ICCCN*, 2018.

[4] Q. P. He and J. Wang, Statistics pattern analysis: A new process monitoring framework and its application to semiconductor batch processes, *AIChE J.*, 2011.

[5] J. Wang and Q. P. He, Multivariate Statistical Process Monitoring Based on Statistics Pattern Analysis, *Ind. Eng. Chem. Res.*, 2010.

633025 Gradient-Weighted Class Activation Mapping (Grad-CAM) Based Explanations for Process Monitoring Results from Deep Neural Networks

Abstract:

In the chemical process industries, high product quality with maximum efficiency and process safety are the critical issues that require effective process monitoring. A fault in a process results in the deviation of process state variables from its normal operating condition. The effect of the fault propagates along the process and disturbs stability of the system. So, it is essential to detect them quickly on their occurrence. To tackle this problem, we motivate to develop the methodology which fulfils the aspect of high accuracy, novel explainability and to make effective use of the multivariate time series process data.

Recent developments in sensor techniques and storage technologies have provided a significant technology push for developing next-generation data driven process monitoring methods [1]. Hence deep learning methods have attracted much attention for fault detection. Deep learning methods like Deep Belief Network (DBN), Convolutional Neural Networks (CNN) and, Hierarchical Deep Neural Networks (HDNN) have been used for fault detection [2] but they lack in providing explanations for their decision. Hence, the eXplainable Artificial Intelligence (XAI) method Integrated Gradients was used by [3] to provide explanations for the results generated from deep learning models. But the Integrated Gradient suffer from one shortcoming that it analyses the samples one by one without accounting for the fact that the data is a time series. In the paper, we develop the XAI method which accounts for the time series nature of the process monitoring problem.

Recently, a new XAI method called Grad-CAM has been developed targeted at providing explanations for time series data in terms of activation maps for CNN [4]. Grad-CAM helps to explore the spatial information that is preserved inside convolution layers. The extracted information from CNN layers helps to understand the important feature of an input data to make classification decision. Grad-CAM uses k feature maps of height h and width u of the last convolution layer, as it contains high level semantics and detailed spatial information. For each feature map A^k, we obtain an importance weight w_k^c associated with specific class c. The w_k^c is obtained by computing the gradient of final class output y^c with respect to the feature map A^k i.e., $\partial y^c / \partial A^k$. It is further globally averaged over width u and height h. This step is followed by computing a weighted sum of feature maps A^k of class c using w_k^c. The final heatmap is obtained by applying the $ReLU$ operation on the weighted sum to emphasize only positive values. This heatmap highlights the features which are important in making a final decision of the network for class c. In this paper, we use the Grad-CAM method for the explanation of process fault diagnosis.

In this paper, we exploit the real time explainable CNN methodology that classifies the normal and faulty state of the process. The proposed methodology consists of a 2D Convolution layer that is trained to solve a fault detection problem. Then, Grad-CAM is used on a trained model to understand which part of input data contributes more to detect a fault [5]. The contribution of different features visualized using the generated heatmap for each sample helps to identify the key factors in fault occurrence. The performance of the proposed methodology demonstrated on simulated time series classification datasets which contain different input variables and categorical outputs. The CNN was trained and heatmaps were generated using Grad-CAM that provide an accurate explanation about the predicted outcome. In this paper, we will describe the proposed methodology as well as results obtained from the simulated dataset which effectively makes use of the time series dimension to provide intuitive explanations for outputs. The results from fault diagnosis on the Tennessee Eastman challenge problem will also be demonstrated [6].

References

[1] P. Park, P. D. Marco, H. Shin, and J. Bang, "Fault Detection and Diagnosis Using Combined Autoencoder and Long Short-Term Memory Network," *Sensors*, vol. 19, no. 21, p. 4612, Oct. 2019, doi: 10.3390/s19214612.

[2] H. Wu and J. Zhao, "Deep convolutional neural network model based chemical process fault diagnosis," *Comput. Chem. Eng.*, vol. 115, pp. 185–197, Jul. 2018, doi: 10.1016/j.compchemeng.2018.04.009.

[3] V. Pakkiriswamy and R. Srinivasan, "An Explainable Artificial Intelligence Based Approach for Interpretation of Fault Detection Results from Deep Neural Networks," presented at the 2020 Virtual AIChE Annual Meeting, 2020, [Online]. Available: https://www.aiche.org/academy/conferences/aiche-annual-meeting/2020/proceeding/paper/375g-explainable-artificial-intelligence-based-approach-interpretation-fault-detection-results-deep.

[4] R. R. Selvaraju, M. Cogswell, A. Das, R. Vedantam, D. Parikh, and D. Batra, "Grad-CAM: Visual Explanations from Deep Networks via Gradient-based Localization," *ArXiv161002391 Cs*, Dec. 2019, doi: 10.1007/s11263-019-01228-7.

[5] R. Assaf and A. Schumann, "Explainable Deep Neural Networks for Multivariate Time Series Predictions," in *Proceedings of the Twenty-Eighth International Joint Conference on Artificial Intelligence*, Macao, China, Aug. 2019, pp. 6488–6490, doi: 10.24963/ijcai.2019/932.

[6] J. J. Downs and E. F. Vogel, "A plant-wide industrial process control problem," *Comput. Chem. Eng.*, vol. 17, no. 3, pp. 245–255, Mar. 1993, doi: 10.1016/0098-1354(93)80018-I.

630074 A Deep Learning Vision System for Classification of Manufacturing Defects

Abstract:

Manufacturing lines for the assembly of pharmaceutical autoinjectors are highly automated and rely on visual inspection stations to efficiently and safely process a large volume of syringes. As pre-filled syringes enter the inspection line, there are a variety of subtle defects in the flange (e.g., chips, cracks) as well as other processing flaws (e.g., missing components) that can make the assembly unfit for use or otherwise problematic for continued processing. Currently, there is a highly-engineered inspection algorithm that is manually tuned for identifying defects from camera images. The existing approach is able to robustly detect important defects, however, a relatively large percent of rejected products are in fact false positives (e.g., cosmetic defects, lighting irregularities). In this work, we have developed a convolutional neural network approach for classifying defects in syringes with comparable sensitivity and higher selectivity than the previous algorithm. We found that preprocessing and noise reduction were critical for training meaningful models from available images. Additionally, we highlight a series of infrastructure challenges that were overcome during image acquisition and algorithm deployment to transfer our work to and from a good manufacturing process (GMP) inspection line. We also illustrate the importance of algorithm-architecture co-design when developing machine learning models in order for the solution to be practically deployed on the available edge computing device (i.e., the vision processor). This work highlights the use of machine learning in conjunction with traditional computer vision to classify defects beyond what can be identified by expert intuition.

628660 Advanced Formulation of Co-Processed Ionic Liquid Drugs *Via* Spray Drying for Incorporation into Solid Dosage Forms

Abstract:

Transforming active pharmaceutical ingredients (APIs) into analogous ionic liquid forms (API-ILs) holds the potential to concurrently solve the increasingly commonplace problems of poor aqueous solubility and solid state instability of APIs. However, as this class of compounds mostly exist as viscous oils or waxes, handling, processing, and formulation are extremely challenging. Therefore, before these promising forms of active pharmaceutical ingredients (APIs) can be routinely incorporated into oral solid dosage forms (OSDs), questions remain as to how best to overcome their particularly poor material properties.

To address this, spray drying API-ILs with other excipients was found to be an excellent solidification method. This can be achieved in one of two ways by the appropriate selection of carrier material to form single or two-phase solid forms and was applicable to a range of API-ILs. Selecting an API-IL insoluble polymer (ethyl cellulose) to form two-phase systems allows for up to 75% w/w API-IL to be loaded into the polymer and recovered as a fine powder, whilst up to 90% w/w API-IL content can be recovered as a solidified paste. Alternatively, an exceptionally high glass transition temperature (T_g) polysaccharide (maltodextrin) can be used to form a single-phase solid solution with the API-IL. Using this approach, however, only allowed for 50% w/w loading of the API-IL in the polysaccharide to be successfully produced. This was due to T_g suppression becoming too substantial at higher loadings for a viable product to be recovered. Perhaps counterintuitively, this points towards the immiscible approach being more desirable for solidification of API-ILs.

As the intended use for the spray dried materials is for incorporation into OSDs, dissolution studies were performed in representative dissolution media and showed that solidification of the API-IL in this manner had no detrimental effect on release characteristics, even when encapsulated in the immiscible (and water insoluble) polymer. The nature of the interactions between API-IL and carrier material were verified using modulated differential scanning calorimetry, hot-stage microscopy, powder X-ray diffraction, and attenuated total reflectance Fourier transform infrared spectroscopy.

These results demonstrate that API-ILs spray dried at high loadings in immiscible or high T_g materials preserve solubility and solid state stability enhancements of the API-IL form. At the same time, these approaches provide solid powders for processing and further formulation unit operations, and so represent an exciting new platform that allows API-ILs to fulfil their potential for formulating poorly soluble compounds as OSDs.

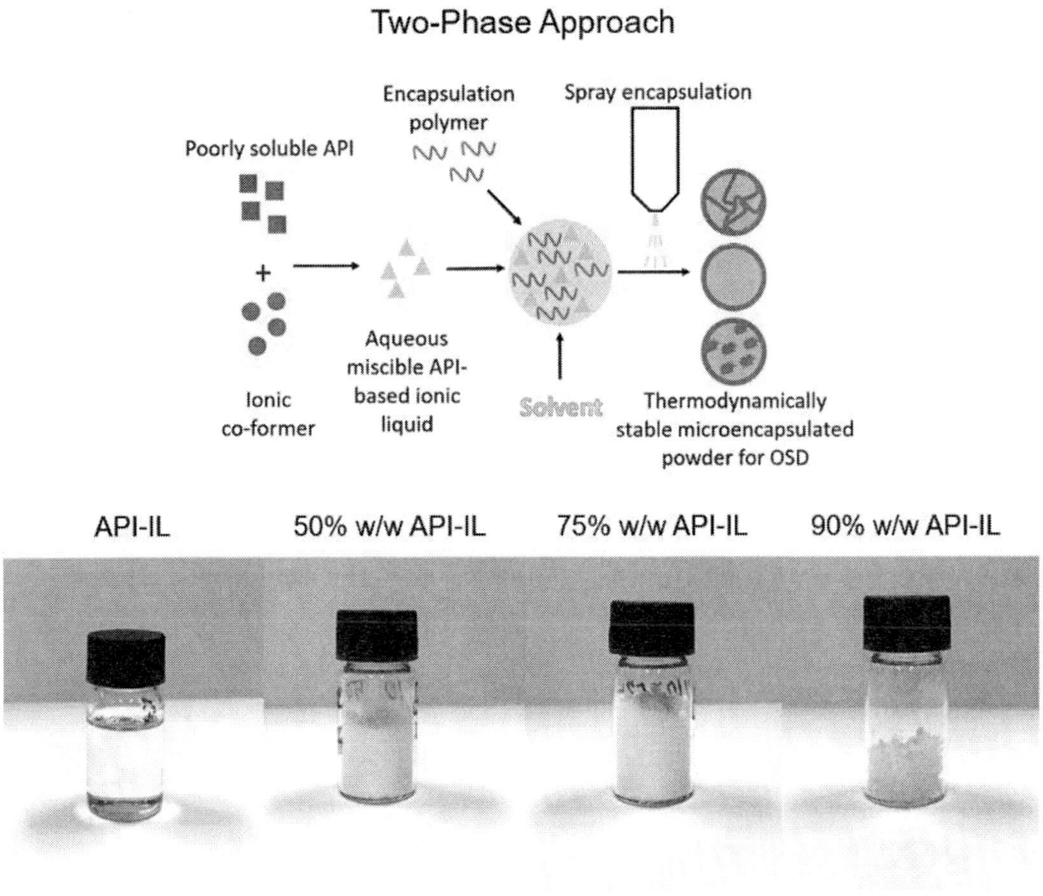
Two-Phase Approach
Encapsulation polymer
Spray encapsulation
Poorly soluble API
+
Aqueous miscible API-based ionic liquid
Ionic co-former
Solvent
Thermodynamically stable microencapsulated powder for OSD
API-IL
50% w/w API-IL
75% w/w API-IL
90% w/w API-IL

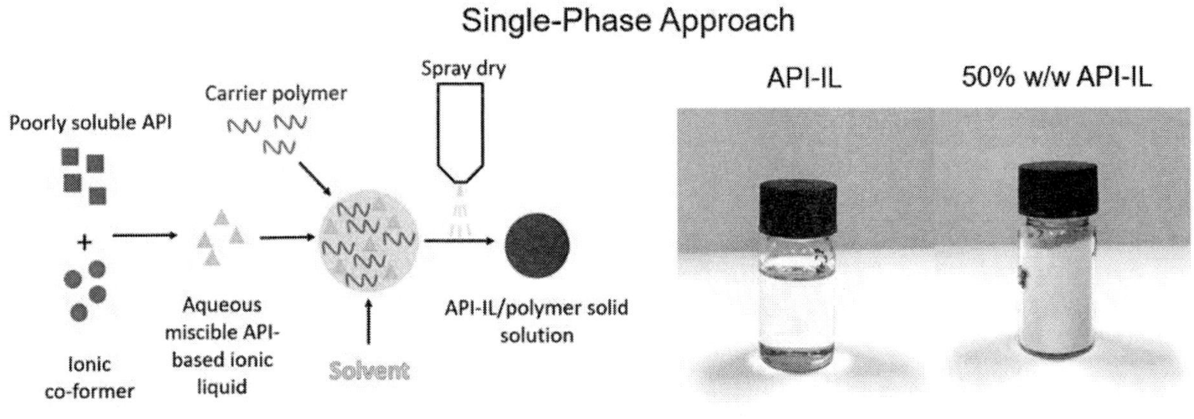
Single-Phase Approach
Spray dry
Carrier polymer
Poorly soluble API
+
Aqueous miscible API-based ionic liquid
Ionic co-former
Solvent
API-IL/polymer solid solution
API-IL
50% w/w API-IL

632246 Rapid Manufacturing Route Conversion Based on the Role of Shear and the Effect of Shear on the Formulation

Abstract:

The current practice, when transferring between batch manufacturing and continuous manufacturing, is to perform multiple designed experiments and perform statistical analysis on the results. While scientifically beneficial, practically this leads to significant waste in material with little to no actionable items. When we take a step back and start comparing between continuous and batch manufacturing, the tablet presses are the same for both manufacturing platforms. Thus, we can assume the table press will impart the same amount of conditioning (shear) to the powder blend in order to create the tablets. Therefore, if the material entering the tablet press has the same properties (bulk packing and flow) for both batch and continuous manufacturing, similar tablet properties should be obtained for both processes. If this is true, the focus of the work should be to identify if the bulk properties of the intermediate blend from one process is essentially equivalent to another process.

Formulations will inevitably end up in two categories. The first is a blend that is insensitive to the conditions at which it is produced, i.e. shear insensitive. The second category are blends that will undergo property changes due to the conditions at which they are made, i.e. shear sensitive. Within the second category, formulations can fall within subcategories, i.e. shear sensitive blends that do not effect the final product, shear sensitive blends that can be manipulated so that the final products can be matched, and shear sensitive blends that will always result in a different final product.

In this project multiple formulations were used to study how the bulk properties of each formulation would change based on increasing levels of shear. Since shear can effect API agglomerate size, lubrication, and granule size, the formulations utilized in this work varied each of these mechanisms and was able to identify shear sensitive and shear insensitive formulations. Next shear sensitive formulations were generated using different manufacturing routes and critical process parameters to identify if the blend can be manipulated enough to generate blends with similar bulk properties. For the batch manufacturing route, a V-blender was utilized to mix the blend at 50, 250, and 500 total revolutions. Next each of the blend was passed through a feedframe at 20, 60, or 100 RPM. Two continuous manufacturing routes were utilized LIW Feeders to dispense the individual components into a continuous blender, at three levels of RPM, and then into a feedframe with three levels of RPM. Two different continuous blenders and feedframes were utilized to represent two distinct manufacturing lines. The three manufacturing routes are identified in Figure 1. All the blends were characterized and blends were identified with the same bulk properties while being produced by different manufacturing routes. Tablets were generated and characterized to show the final products were in-fact equivalent. The results indicate that the critical process parameters of the manufacturing route can be optimized in order for the bulk blend properties can be matched independent of manufacturing route. No extensive designed experimental plan would be required to operate the manufacturing route at conditions that are unlikely to ever be utilized at the manufacturing facility. By matching blend properties, time, material, and money can be saved when converting formulations between manufacturing routes.

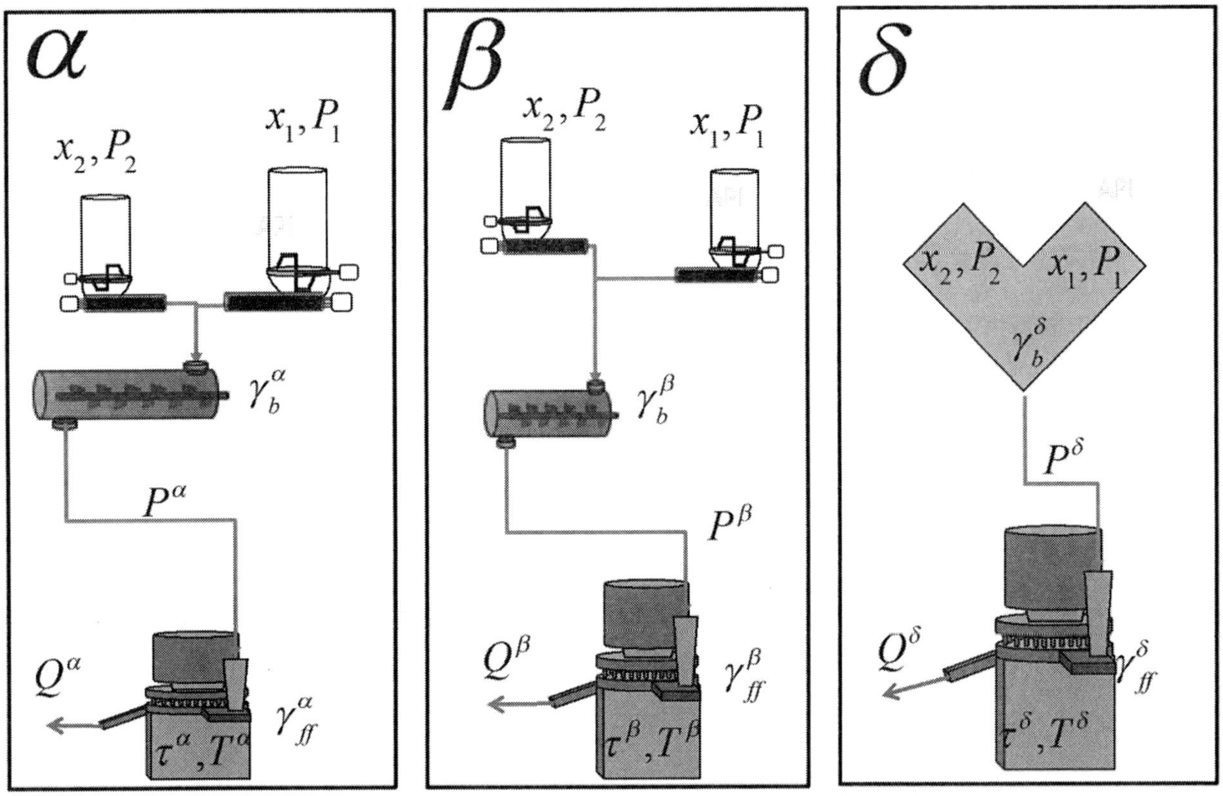

Figure 1: the figure displays a first continuous line α, a second continuous line β, and a batch line δ, all used to make the same product.

624533 An *in silico* Tool for Quantitative Kinetic Predictions of API Degradation

Abstract:

Pharmaceuticals are susceptible to oxidation and hydrolysis that often lead to potency loss and the formation of harmful impurities. To ensure drug product efficacy and quality, it is critical to elucidate degradation pathways and products of active pharmaceutical ingredients (APIs). Contemporary approaches to studying API stability rely on forced API degradation experiments. In a typical API degradation study, researchers conjecture API degradation pathways and products at first and then verify the hypotheses experimentally. This trial-and-error approach is inefficient as API degradation experiments are time-consuming and common analytic methods (e.g., HPLC-MS, NMR) consume significant amounts of API, which is expensive to synthesize and purify especially at early stages of drug R&D. Moreover, researchers often have limited intuition of what degradants to expect for a new API, making HPLC-MS analyses challenging.

Alternatively, supplementing the contemporary experimental approach with novel computational chemistry methods has the potential to significantly accelerate API stability studies and reduce R&D costs. **In this talk, we present the development of a self-improving software that can readily predict API degradants, automatically build API degradation kinetic models, and reliably identify critical API decomposition pathways.** Pfizer and MIT researchers collaboratively investigated the aqueous oxidative degradation chemistry of imipramine. The imipramine degradation model constructed by the MIT team using the newly developed software successfully predicted degradants observed experimentally by the Pfizer team. The model and experimental results also provide novel insights into imipramine degradation pathways and kinetics.

Based on the promising results, we believe the quantitative *in silico* tools that we are actively developing will greatly assist the pharmaceutical research community in performing drug stability studies. The computer-assisted drug stability analyses will guide future drug degradation experiments, assist in API formulation improvements to resist oxidation, and potentially reduce drug development costs.

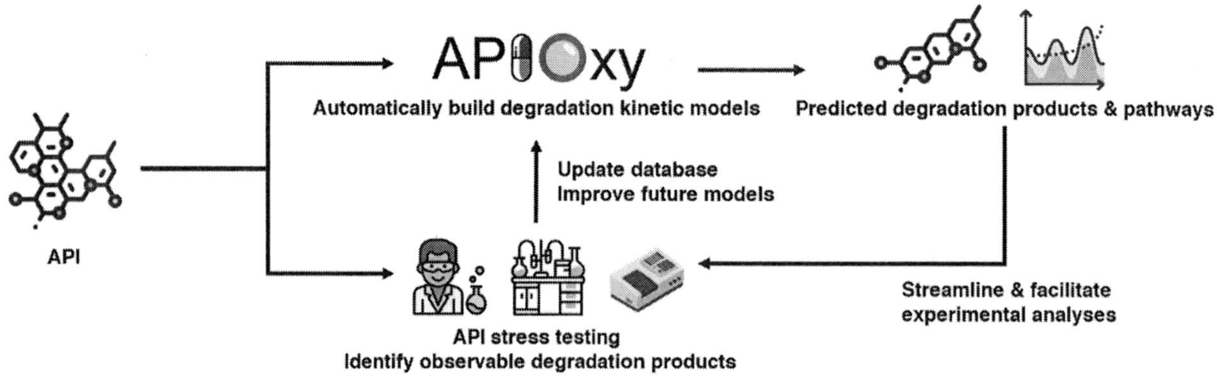

630691 Structure - Function Analysis of Lipids for the Development of Advanced Excipients for Pharmaceutical Manufacturing

Abstract:

Background and objective

The gradually increased expenses of global pharmaceutical R&D can be explained by: i) the rising number of new FDA approvals in the recent years, ii) increased awareness of pharmaceutical companies of the sustainability of their production (1) and iii) increased awareness of the specific requirements of patients as individuals (2,3). These trends are closely associated with the demand for innovative and robust dosage forms with stable performance. Using excipients with desired functionality is the key parameter for realizing this demand. In this context, lipid-based excipients (LBEs) have been increasingly applied for several reasons. LBEs are naturally occurring materials that are predominantly digestible and have a "Generally Recognized as Safe" (GRAS) status. They are low-toxic, biocompatible, easily available and they are used for manufacturing a broad range of pharmaceutical dosage forms. The serious challenge, however, is their unstable solid state, which affects the stability and robustness of pharmaceutical products.

In this work, the ideal properties of LBE are defined from molecular to macroscopic level. Following this definition, polyglycerol fatty acid esters (PGFAs) available as Witepsol® PMFs are introduced as such LBEs and their structure-function relationship are analyzed with some examples for their application as advanced LBE in the development of pharmaceutical dosage forms.

Ideal properties of LBEs can be defined as:

- Direct crystallization or a fast transformation into the most stable polymorphic form (no polymorphism).
- Being a one-phase system to avoid phase separation and crystalline growth during the storage.
- Diversity in the functional groups of the chemical structure in order to offer diversity in the physicochemical properties in the macrostructure level. This limits the need for additives to the formulation and thus limits the crystal growth and alteration in the crystalline network.

Polyglycerol fatty acid esters (PGFAs) and their structure-function relationship

It is known that a polymorphic change, for example in mono- di- or triacylglycerols, is due to intermolecular non-covalent interactions between hydrocarbon chains, causing the sub-cell rearrangement toward the most thermodynamically stable geometry. Therefore, providing enough space between the chains can be a reasonable way to avoid polymorphic transition. In this context, polyglycerol esters of fatty acids (PGFAs) are promising molecules.

PGFAs are hydroxyethers of glycerol, fully or partially esterified with fatty acids (Figure 1). The larger space among hydrocarbon chains of a PGFA molecule, which is caused by the ether bonds connecting the glycerol moieties, can weaken the intermolecular interactions between the chains and thus avoid a polymorphic transformation. Analyzing the thermal behavior of selected PGFAs as received and after 6 months storage at 40 °C via X-ray diffraction and DSC revealed the crystallization of all samples into the stable monophasic α-form without transformation during storage (4-6).

Tuning the number of glycerol moieties, fatty acids chain length and free hydroxyl groups per molecule resulted in different hydrophilic-lipophilic balance (HLB) values, wettabilities, melting points, and melt viscosities (4). Structure-function relationship analysis of PGFAs showed that an increase in available free hydroxyl groups results in shifting to smaller, less dense and more disordered crystal clusters. The HLB value was also increased by increasing the number of glycerol moieties and/or available number of free hydroxyl groups (partial esterification). As expected, the HLB value was directly correlated with the wettability of molecules. An increase in the number of glycerol moieties and thereby increased number of hydroxyl groups also resulted in the increased melt viscosity of PGFAs, which is due to the lower mobility of the long chains and the availability of polar groups for hydrogen bonding. The melting temperature was directly correlated with the length of the fatty acid chain.

Application of PGFAs in the development of pharmaceutical dosage forms

Selected compounds were used as hot melt coating (HMC) excipients (5), for manufacturing of solid lipid nanosuspensions (SLN) (6), spray-dried particles for inhalation as well as for cold extrusion.

Hot melt coating: The HMC trials were performed in a fluid bed equipment. N-acetylcysteine was coated with three PGFAs with different HLB values: PG3C16/C18-partial, PG4C18-partial and PG6C18-partial (Witepsol® PMF 1683, 184, and 186). The DSC data showed the absence of low solidification fractions, thus reduced risk of agglomeration during the coating process. The driving force for crystallization was lower and the heat flow exotherms were broader compared to conventional HMC formulations, indicating a lower energy barrier for nucleation. The API release was directly proportional to HLB and was stable during storage (5).

Manufacturing of SLN: SLN were produced via melt-emulsification followed by high pressure homogenization using PG2C18-full (Witepsol® PMF 282) as lipid matrix and poloxamer 188 as stabilizer to encapsulate the model API dexamethasone. SLN with an API loading of 0.1%, a sharp median particle size of 242.1 ± 12.4 nm, a zeta potential of -28.5 ± 7.8 mV, an entrapment efficiency (EE) of 90.2 ± 0.7% were produced. The crystallization of the lipid matrix in the stable α-form was not affected by process conditions or material interactions. Stable EE and release profile during storage was observed. The pulmonary *in vitro* safety of SLN was given (6).

Manufacturing of spray-dried particles for inhalation: Lipid-microparticles for systemic delivery were loaded with ibuprofen under an organic solvent-based spray drying process. PG3C22-partial (Witepsol® PMF 123) was used as the lipid matrix in a lipid:drug ratio of 70:30. The process resulted in high yield of solid particles (71%). Particles with unimodal size distribution, volume mean diameter (VMD) of 6.6 ± 1.1 µm and density of 0.389 ± 0.007 g/cm3 were achieved. The mass median aerodynamic diameter (MMAD) of particles was 3.57 ± 0.11 µm, indicating the inhalability of the particles. The fine particle fraction (FPF) was 45.2 ± 1.6%. Dissolution of the deposited particles in simulated lung fluid (SLF) showed stable modified release within 6 hours.

Cold extrusion: PG6C16-partial and PG2C18-full (Witepsol PMF166 and 282, respectively) were processed via cold extrusion at temperatures below their melting onset. Both materials were extrudable, resulting in filaments without die swell and a homogeneous diameter distribution. Filaments of PG6C16-partial showed improved elasticity and spoolability compared to PG2C18-full, which can be explained by higher number of glycerol moieties. These filaments can be either processed into multiparticulate systems in a further down-streaming step or can be used for printing of solid dosage forms via extrusion-based 3D-printing. Loading of API in the filaments is currently under investigation.

Conclusion

The stable release profile of API from HMC particles, SLN, and spray-dried powder for inhalation, and the stable particle size distribution and EE of SLN over storage were attributed to the monophasic crystallization of PGFAs into stable α-form and no crystallite growth. In all case studies the release profile could be tailored based on the HLB of selected PGFAs. In case of SLN, the application of PGFAs as stable lipid matrices, without the addition of oils, can pave the way for developing next generation SLN. The extrudability of PGFAs in flexible filaments plays an essential role in the extension of LBE application in pharmaceutical drug development.

Acknowledgement

The Austrian Research Promotion Agency (FFG)

References

1. Cue, B. W., & Zhang, J. (2009), Green Chemistry Letters and Reviews, 2(4), 193–211. https://doi.org/10.1080/17518250903258150
2. Penkov, D., et al., (2017), Therapeutic Innovation and Regulatory Science, 51(3), 360–371. https://doi.org/10.1177/2168479017696265
3. Stegemann, S. (2016), International Journal of Pharmaceutics, 512(2), 334–342. https://doi.org/10.1016/j.ijpharm.2016.01.051
4. Corzo C. et al., (2020), European Journal of Pharmaceutics and Biopharmaceutics, 148, 134–147. https://doi.org/10.1016/j.ejpb.2020.01.012
5. Salar-Behzadi S., et al., (2020), European Journal of Pharmaceutics and Biopharmaceutics, 148, 107–117. https://doi.org/10.1016/j.ejpb.2020.01.0093.
6. Corzo C. et al., (2020), European Journal of Pharmaceutics and Biopharmaceutics, 152, 44–55. https://doi.org/10.1016/j.ejpb.2020.04.027

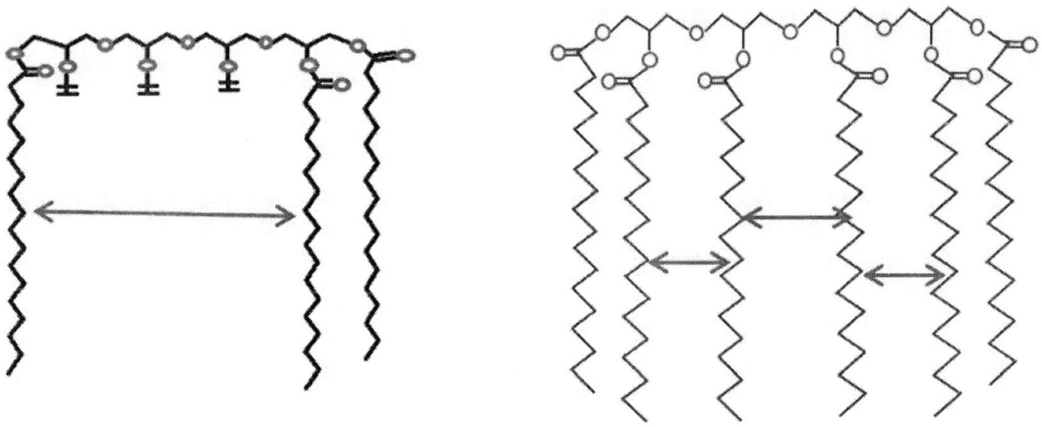

Figure 1- Chemical structure of PGFAs as hydroxyethers of glycerol, fully (right-hand) or partially (left-hand) esterified with fatty acids.

630482 Implementation of a Methodology for Selection of an Appropriate Tracer to Measure the Residence Time Distribution (RTD) of Continuous Powder Blending Operations

Abstract:

Residence time distribution (RTD) is used as a tool to measure the degree of back-mixing by determining the time (mean and distribution) a particle resides in a unit operation [1]. The RTD is commonly measured experimentally by tracer pulse injection into the unit operation entrance at time t=0, and then the tracer concentration is measured, using process analytical technology (PAT), in the exit as a function of time [2]. Choosing an appropriate tracer is of great importance. It is well accepted that a tracer should be detectable, non-interactive with the system, have the same physical properties similar to those of the bulk particles, be able to mix with other system components, and not adhere to equipment surfaces [3, 4]. However, for powder systems, accomplishing all these requirements may prove challenging.

It appears that the proper selection of tracers has not received sufficient attention in recent efforts to characterize the RTD of continuous pharmaceutical powder processes. Tracers are often selected *ad-hoc* or based on the availability of materials with easy means of detection. Although this is an essential requirement for tracer materials (i.e., APIs often have distinct spectral peaks), it is equally important to meet other tracer properties' requirements to avoid misinterpretations of mixing behavior inside the unit. Otherwise, any models based on such results will likely fail to predict materials' evolution through the system and could have significant consequences such as in control decisions for conforming/non-conforming intermediate material and final product. Thus, an accurate measure of RTDs, a large part of which is the proper selection of tracers, is of utmost importance.

We provide a methodology to choose a tracer that can best describe the dynamic behavior of the blend in the unit. First, the material properties of the API need to be determined using a standard suite of measurements, including particle size distribution, bulk and tapped density, and flow properties. Next, the API material properties are compared to all materials in a material property database, containing over 150 commercially available materials. The comparison method is based on determining the weighted Euclidean distance in principal component space. Materials with similar properties to the API are identified and ranked for suitability to use as tracers (e.g., being detectable). Lastly, the similarity of the RTD of the API and the tracer candidate needs to be verified. To this end, under steady-state conditions, the RTD is obtained by injecting (1) a pulse of the API and (2) a pulse of the tracer.

In this study, the problem was simplified using one excipient as the base powder. However, since formulation flow properties are affected by the API concentration and the rest of the ingredients properties, the experiments were repeated for three different excipients varying in their flow behavior applicable to the continuous direct compression line. The outcome of these experiments enables the side-by-side comparison of RTDs obtained using pulses of either API or tracer for powders with a varying range of properties. Moreover, the effect of tracer amount on RTD was explored to identify an appropriate amount that is small so that it does not perturb the system yet enough to be detected.

References:

1. Engisch, W. and F. Muzzio, *Using residence time distributions (RTDs) to address the*

traceability of raw materials in continuous pharmaceutical manufacturing. Journal of pharmaceutical innovation, 2016. **11**(1): p. 64-81.

2. Levenspiel, O., *Chemical reaction engineering.* 1999: John Wiley & Sons.
3. Escotet-Espinoza, M.S., et al., *Effect of tracer material properties on the residence time distribution (RTD) of continuous powder blending operations. Part I of II: Experimental evaluation.* Powder Technology, 2019. **342**: p. 744-763.
4. Danckwerts, P.V., *Continuous flow systems: distribution of residence times.* Chemical engineering science, 1953. **2**(1): p. 1-13.

623267 Development of Discrete Element Method Calibration Approach for Pharmaceutical Applications

Abstract:

Discrete element modeling (DEM) has been extensively used for detailed simulation of particulate physics[1]across various industries. In pharmaceutical manufacturing, DEM modeling has also been widely applied for various applications[2,3] like material transport and storage, feeding, blending, granulation, milling, compression and film coating, to obtain a detailed insight into the powder flow and powder dynamics. Although DEM has immense advantages for simulating particulate flow, there exists significant limitations for such applications. One of the critical limitations is the accurate calibration of the powder system in DEM. This corresponds to evaluating the DEM parameters which best represent the powder in DEM space, as observed in an experimental setting. Calibration is a widely studied problem in DEM literature, and researchers have shown different approaches using bulk properties measurements to handle these issues [4-6]. However, most of these studies[5,7] available in the literature are limited to specific powders or specific processing conditions and thus, the calibration process needs to be repeated every time a powder or processing condition is encountered. Furthermore, the literature lacks in addressing the problem of solution multiplicity for the problem of DEM calibration[8,9], which states that there can be multiple possible solutions that replicate the bulk behavior, which makes it difficult to identify a unique set of DEM parameters.

In the proposed work, the authors aim to address the problem of DEM calibration for pharmaceutical manufacturing using a unique strategy utilizing a combination of multiple bulk measurement tests. The combination of bulk measurement tests includes shear cell test, FT4 flow energy test and an instrumented rotating drum to account for static and consolidated state, dynamic flow regime and non-consolidated state of powder flow respectively. These tests aim to span the majority regime of observed powder flow behavior and processing conditions in different applications for pharmaceutical manufacturing. Following the demonstration of material calibration, the proposed work aims at extending the idea to develop a novel concept of DEM calibration database. The DEM calibration database, equivalent to the material characterization library, includes the calibrated DEM parameters of commonly used pharmaceutical powders, selected from different clusters of the material library[10]. Multivariate analysis techniques like principal component analysis and clustering analysis are implemented to explore the knowledge space of the database. Lastly, the DEM calibration database is validated for new pharmaceutical powders based on predictive models constructed using surrogate modeling. The novelty of the proposed work is that the developed DEM calibration space can then be used as a lookup guide for quick access to calibrated DEM parameters of known pharmaceutical powders.

References:

[1] P.A. Cundall, O.D.L. Strack, A discrete numerical model for granular assemblies, Géotechnique. 30 (1980) 331–336. doi:10.1680/geot.1980.30.3.331.

[2] W.R. Ketterhagen, M.T.A. Ende, B.C. Hancock, Process modeling in the pharmaceutical industry using the discrete element method, Journal of Pharmaceutical Sciences. 98 (2009) 442–470. doi:10.1002/jps.21466.

[3] S. Bin Yeom, E.-S. Ha, M.-S. Kim, S.H. Jeong, S.-J. Hwang, Du Hyung Choi, Application of the

Discrete Element Method for Manufacturing Process Simulation in the Pharmaceutical Industry, Pharmaceutics 2020, Vol. 12, Page 235. 11 (2019) 414. doi:10.3390/pharmaceutics11080414.

[4] C. Coetzee, Calibration of the discrete element method: Strategies for spherical and non-spherical particles, Powder Technology. 364 (2020) 851–878. doi:10.1016/j.powtec.2020.01.076.

[5] C.J. Coetzee, Review: Calibration of the discrete element method, Powder Technology. 310 (2017) 104–142. doi:10.1016/j.powtec.2017.01.015.

[6] M.W. Johnstone, Calibration of DEM models for granular materials using bulk physical tests, The University of Edinburgh, 2010.

[7] M. Marigo, E.H. Stitt, Discrete Element Method (DEM) for Industrial Applications: Comments on Calibration and Validation for the Modelling of Cylindrical Pellets, KONA Powder and Particle Journal. 32 (2015) 236–252. doi:10.14356/kona.2015016.

[8] T. Roessler, C. Richter, A. Katterfeld, F. Will, Development of a standard calibration procedure for the DEM parameters of cohesionless bulk materials – part I: Solving the problem of ambiguous parameter combinations, Powder Technology. 343 (2019) 803–812. doi:10.1016/j.powtec.2018.11.034.

[9] C. Richter, T. Robler, G. Kunze, A. Katterfeld, F. Will, Development of a standard calibration procedure for the DEM parameters of cohesionless bulk materials – Part II: Efficient optimization-based calibration, Powder Technology. 360 (2020) 967–976. doi:10.1016/j.powtec.2019.10.052.

[10] M.S. Escotet-Espinoza, S. Moghtadernejad, J. Scicolone, Y. Wang, G. Pereira, E. Schafer, et al., Using a material property library to find surrogate materials for pharmaceutical process development, Powder Technology. 339 (2018) 659–676. doi:10.1016/j.powtec.2018.08.042.

630183 High-Efficient Crystal Particle Manufacture Via Microscale Process Intensification Technology

Abstract:

High-end crystal manufacture has drawn a permanent concern on the high-efficient manufacture of crystal particles, especially in fine chemical, pharmaceutical, electronics, biological and relative engineering fields. In recent years, a series of micro-scale process intensification (MPI) technologies have been widely used in crystal particles preparation via addressing the control of nucleation and growth process. Herein, we review the research progresses of microscale process intensification technology from three aspects, microfluidics devices, microscale force field technology and membrane-based microchannels and interface transfer process. Firstly, the principle of microfluidic and relative microscale device on improving micro-mixing and mass transfer are briefly described. The advantage of microfluidic in continuous nano particle preparation is outlined. Microscale external force field (ultrasonic, high-gravity, electric and magnetic fields) is then introduced as another novel approach for ultrafine nanoparticles and continuous drug crystallization process. Further, in view of the micro-scale intensified mass transfer and micro-scale interfacial force field established on membrane technology, the basic mechanism of membrane crystallization (microscale 2D supersaturation degree control, auto seed detachment, microporous membrane dispersion, etc.) are reviewed. The process coupling and design strategy aiming for enhancing the manufacture capacity is also illustrated. Finally, the developing tendency and key challenges of high-efficient crystal particle preparation technology via microscale processes are overviewed.

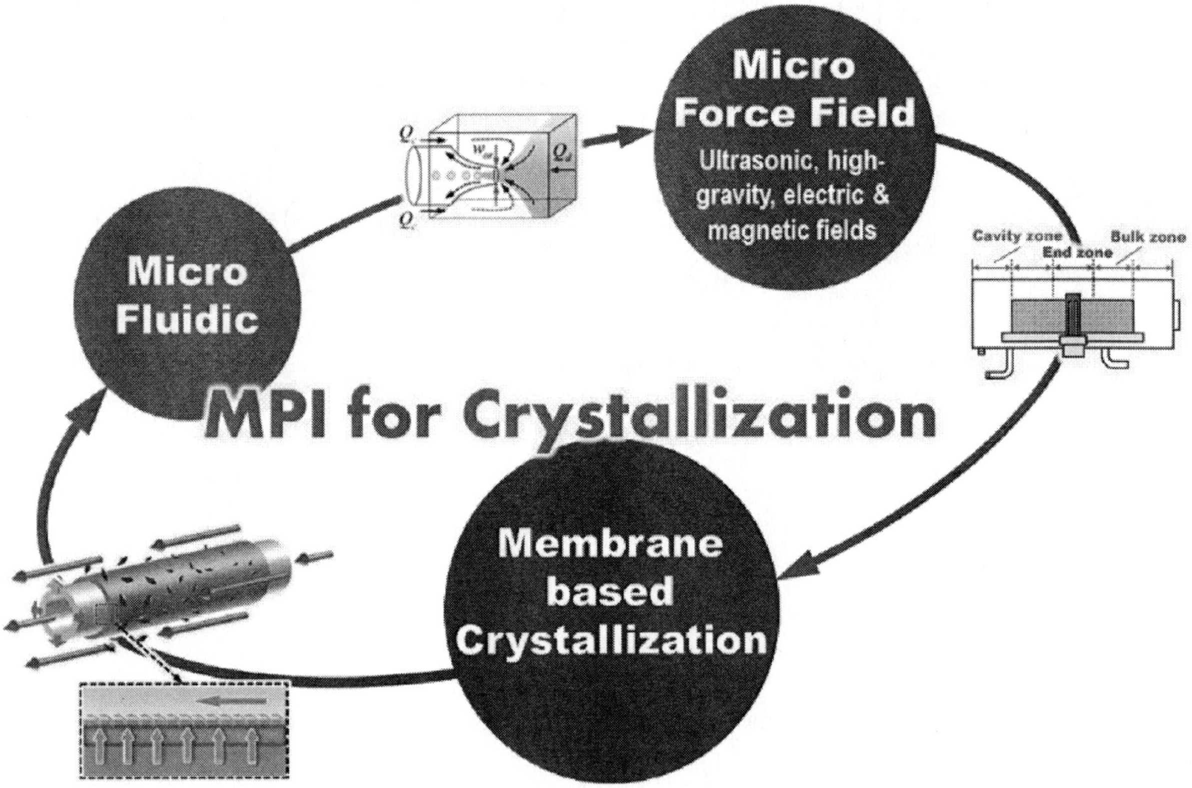

623376 Breakage Facilitated Mixed-Suspension, Mixed-Product-Removal (MSMPR) Crystallization

Abstract:

Continuous crystallization provides advantages of producing solid products with desired quality with smaller footprint. However, the intrinsic nucleation and growth rates of small molecule pharmaceuticals constrain the design spaces of the ranges of robust operation conditions and product properties. In this work, a wet-milling device, used to break large particles into small fragments, was incorporated to MSMSR single stage crystallization. The MSMPR crystallization was controlled by a feedback loop in order to maintain a constant total slurry volume. The approach was applied to paracetamol to demonstrate the effectiveness. Experimental data showed that the introduction of the controllable breakage mechanism offered better particle size distribution control and minimized fouling that resulted in longer operation time than conventional MSMPR.

Ref:

Li, H.; Yang, B.-S. Model Evaluation of Particle Breakage Facilitated Process Intensification for Mixed-Suspension-Mixed-Product-Removal (MSMPR) Crystallization. *Chem. Eng. Sci.* 2019, *207*, 1175– 1186, DOI: 10.1016/j.ces.2019.07.030

629001 Electrochemical Sensor-Integrated, Continuous Flow, Microfluidic Device for Real-Time Measurement of Supersaturation during Salt Screening

Abstract:

Electrochemical sensors can offer quick and robust detection of organic molecules inside microfluidic devices where the amount of sample is small. These sensors can be assembled with a feedback control unit for automatic adjustments of flowrates to obtain higher experimental accuracy. The sensor-integrated microfluidic system can facilitate high-throughput screening of crystalline materials in shorter time by precisely controlling the concentration of organic molecules in each microwell. In this study, a continuous microfluidic mixer is coupled with an electrochemical sensor to enable in-situ measurement of solute concentration in the mixing zone. The sensor consists of three electrodes: working, contour, and reference connected externally to a potentiostat. The electrodes are inserted into the microfluidic mixer horizontally. Here we have used Pt and Ni wires as the counter and working electrodes, respectively. Cyclic Voltammetry (CV) is performed during the continuous operation of the microfluidic device which correlates to the concentration of Naproxen. This sensor integrated microfluidic device is experimentally evaluated for the salt screening of Naproxen. Screening of solid salt forms of Naproxen includes four solvents and four salt formers (SFs) which results in sixteen experimental conditions. All experiments are performed under an optical microscope for in-line measurements of crystals' morphology and growth rates. After the concentration measurement and salt formations, crystallized salts are harvested for PXRD and particle size distribution analysis.

623691 Periodic Forcing Via Temperature Cycles: Model-Based Study of Novel Continuous Configurations for Deracemization

Abstract:

Introduction

Crystallization is an attractive approach to achieve targeted production and isolation of the desired enantiomer in a productive and economically effective way due to its low capital and operating costs. Solid-state deracemization techniques, such as solid-state deracemization via temperature cycles can be applied to convert a racemic mixture of crystals to pure enantiomers by the periodic cycling of the temperature and by the utilization of a liquid-phase catalyst [1]. This process is conventionally operated in batch mode and it has been widely investigated in both experimental and simulation works [1,2,3]. However, its continuous counterpart that can be performed in mixed suspension mixed product removal crystallizers (MSMPRC) has not yet been proposed. This novel process can be more attractive than Preferential crystallization, which can reach only 50% yield.

In this work, we will focus on the model-based design and evaluation of four continuous configurations to perform Periodic Forcing via Temperature cycles (PFTC) and we comparatively assess their performance based on the enantiomeric purity and productivity, i.e., the key performance indicators (KPIs).

Method

The mathematical model of the continuous process configurations was obtained by extending our population balance based model of batch temperature cycles [3]. Deracemization via temperature cycles is a complex interplay of several phenomena and mechanisms, thus the description of size-dependent solubility, crystal growth/dissolution, agglomeration, breakage and racemization are incorporated in the model.

Process configurations

Different configurations could be utilized to perform solid-state deracemization via temperature cycles in an MSMPRC. In all these cases, the MSMPRC initially contains a saturated solution, which is in contact with the seed crystals that are enriched in the target enantiomer and temperature cycles are applied periodically during the process. The configurations differ from each other in the type of the feed: in the first case, we feed racemic crystals and we aim at converting them into the desired enantiomer (configuration (A)). In this case, suspension is withdrawn and the solution is recycled after filtration to keep the volume of the crystallizer constant. For this configuration to be more productive and to reach larger enantiomeric excess, one can attempt to create larger asymmetry between the two enantiomers by partially recycling the obtained solid product (configuration (B)). Apart from feeding crystals, one can feed a suspension to the MSMPRC in an alternative configuration (configuration (C)). One could expect that by feeding not only racemic solid, but solution as well, the concentration of the enantiomers becomes more similar to each other as well, thus the driving force for racemization decreases and the deracemization process is hindered. It is worth noting that the only experimental example of continuous temperature cycling process, which employs spatial thermal oscillations in a tubular flow milli-reactor [4], also utilizes a suspension type of feed.

In continuous crystallization, a mechanism is needed to form new crystals, otherwise complete washing out of crystals from the crystallizer occurs. In the three configurations presented above, the feed contained racemic crystals, thus other crystal-generating mechanisms, e.g., attrition, breakage or nucleation, was not essential. However, in the fourth configuration we propose, only solution is fed (configuration (D)). Therefore, for the sake of simplicity, we have introduced breakage due to wet-milling to the system. It is easy to foresee that with this configuration one can attain larger productivity and enantiomeric purity: as a result of introducing the undesired enantiomer into the system in the solution, one step (the dissolution of the undesired enantiomer) can be spared with respect to the solid-fed configuration.

Results and discussion

With the aid of the mathematical model, we have simulated continuous temperature cycles performed in the proposed continuous configurations [5]. We have analyzed first the solid-fed configuration (configuration (A)) in detail: the enantiomeric excess increases during the startup, and then reaches a periodic steady state, which is bounded by the lower and upper limits corresponding to the enantiomeric excess at low and high temperature, respectively. The average periodic steady state enantiomeric excess was chosen as a key performance indicator, together with the average periodic steady state productivity, to compare the processes. A simulation study on the effect of operating conditions, such as the residence time, feed rate and cooling rate was carried out. We have identified that by increasing the residence time or the feed rate, one can significantly improve the process performance.

Second, we have verified that by partial recycling of the crystalline product, slightly higher average steady state enantiomeric excesss and productivity can be attained (configuration (B)). Third, we have investigated the configuration with the suspension feed (configuration (C)). The average steady state enantiomeric excess was found to be lower than that of the solid-fed process, as expected. We have explored the effect of varying the operating parameters to increase the process performance. Then, we have used the model to simulate the solution-fed operation (configuration (D)). Finally, we have studied an MSMPRC cascade for the production of enantiopure crystals.

Conclusions

Various novel MSMPRC-based approaches have been proposed by means of model-based studies to perform solid-state deracemization via temperature cycles in continuous operation. The advantages and pitfalls of each technique have been assessed, together with the potential performance improvements achieved by carefully selecting the operating conditions. This contribution serves as a guide to transform the conventional batch process of temperature cycles to continuous operation.

References

- [1] Suwannasang, A. E. Flood, C. Rougeot, G. Coquerel, Org. Process Res. Dev. 21 (2017) 623-630.
- [2] Breveglieri, G. M. Maggioni, M. Mazzotti, Cryst. Growth Des. 18 (2018) 1873-1881.
- [3] Bodák, G.M. Maggioni, M. Mazzotti, Cryst. Growth Des. 18 (2018) 7122-7131.
- [4] Cameli, C. Xiouras, G. D. Stefanidis, CrystEngComm, 22 (2020) 3519-3525.
- [5] Bodák and M. Mazzotti, in preparation.

Acknowledgements

This research has received funding as part of the CORE project (October 2016 – September 2020) from the European Union's Horizon 2020 research and innovation programme under the Marie Sklodowska-Curie grant agreement No 722456 CORE ITN

This project has received funding from the European Research Council (ERC) under the European Union's Horizon 2020 research and innovation programme under grant agreement No 2-73959-18.

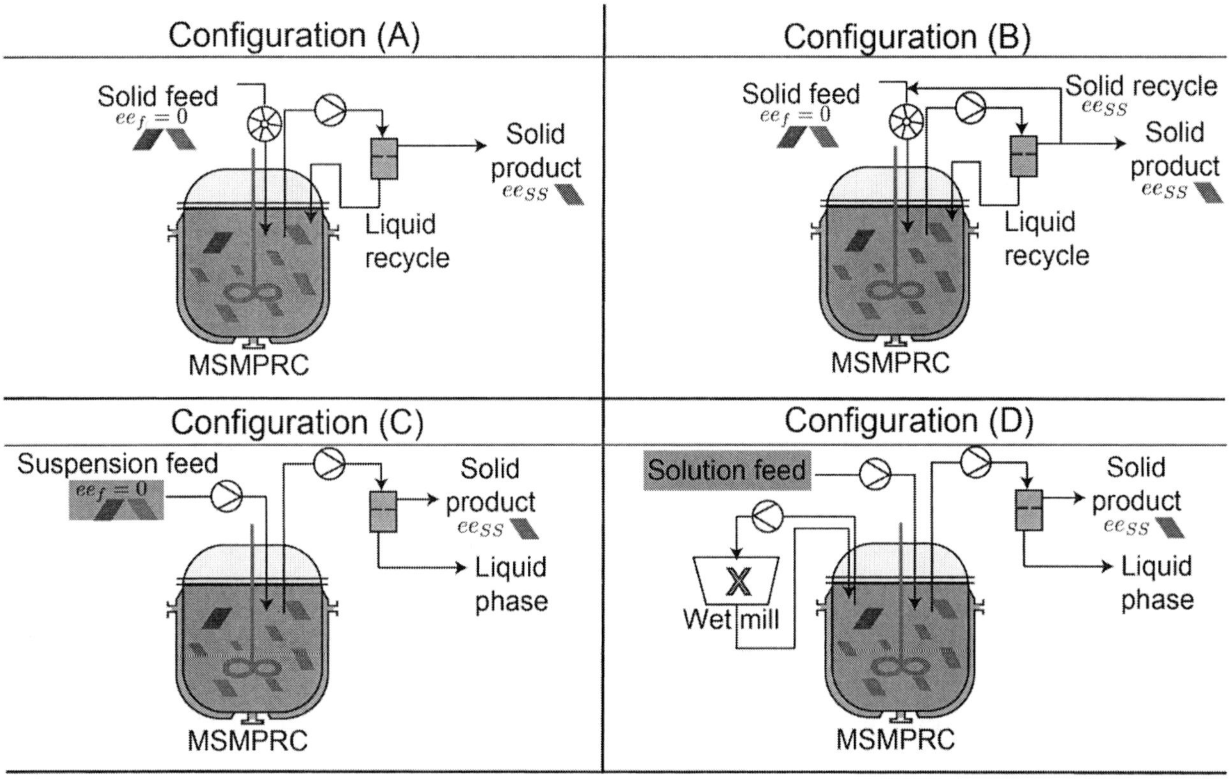

Figure 1. Novel crystallizer configurations are proposed to perform periodic forcing via temperature cycles (PFTC).

621386 Continuous Flow Synthesis of the Metal-Organic Framework HKUST-1 in a Millifluidic Reactor

Abstract:

Metal-organic frameworks (MOFs) are a class of crystalline and porous adsorbents, with wide-ranging applications in gas separations, membrane materials as well as sensors. The basic unit of a MOF consists of metal ions coordinated to organic ligands, which form a coordination polymer with well-defined lattice positions in the crystalline unit cell. The properties of a particular MOF can be finely tuned for a given application through the use of multiple ligands, metal ions and modulators for tuning particle sizes. Conventional lab-scale synthesis procedures for MOFs are autoclave batch reactions with residence times of a day or more. These procedures entail solvothermal synthesis, with expensive and often toxic solvents and starting materials. This explains the reason for a few of them being sold on commercial chemical vendor outlets at exorbitant rates. Thus, given their multifarious applications, economically viable synthesis of MOFs at the industrial scale is the need of the hour.

In this study, we have developed a continuous, droplet-based synthesis procedure of the MOF HKUST-1 in a millifluidic reactor. The droplets consist of aqueous solutions of copper ions, trimesic acid, and various solvent mixtures. Millifluidic reactor comprised of a custom made heating unit including machined metal block and 3D printed parts. Various reaction chemistries for synthesis of HKUST-1 via continuous and batch operation were investigated. We present an ultra-low residence time continuous synthesis of HKUST-1 and compare the physical characteristics and yield with batch synthesis of the same.

631490 Controlling the Effect of Slug-to-Slug Variation on the Crystal Size Distribution of Perovskite QDs: A CFD-Based Approach

Abstract:

Recently, manufacturing of metal halide perovskite quantum dots (QDs), particularly cesium lead bromide ($CsPbBr_3$), has received substantial attention due to its superior tunable optoelectronic properties (related to their size), and because they cater to wide array of applications such as solar cells, next generation displays, and LEDs. [1] Thus, it is of paramount importance to develop a manufacturing route which results in controlled size and size distribution with a desired surface passivation of $CsPbBr_3$ QDs for the targeted application in photonic and optoelectronic devices. Furthermore, device-level applications demand a fast and large scale production of application-ready QDs. However, the large-scale production of these QDs has posed certain engineering and practical challenges. Specifically, an absence of a well-established crystallization model, lack of batch-to-continuous scale-up studies, and unavailability of set-point tracking platforms are the major roadblocks. [2] The fact that slug flow crystallizers (SFCs) offer a large operation range (*i.e.*, precursor concentration, gas-liquid flow rates) has inspired recent studies on proposing SFCs for large scale continuous manufacturing of metal halide perovskite QDs. [3]

Despite the intriguing advantages of SFCs, there are still some key issues with the operation of SFCs. For example, previous studies have not accounted for the effect of slug-to-slug (S2S) variation on crystal size distribution (CSD), and the absence of a modeling and control framework for $CsPbBr_3$ QDs made it challenging to fine-tune the QD size distribution. [4] In response, we have developed a computational fluid dynamics (CFD) model to elucidate the mechanism of slug formation process in a millifluidic SFC, and then combined it with a slug crystallizer model to accurately model QD size distributions and precursor concentration profiles in SFC. Specifically, the slug crystallizer model was constructed by combining a continuum model with a previously developed kinetic Monte Carlo (kMC) model. [5] Based on the CFD-based multiscale modeling framework (*i.e.*, CFD + continuum + kMC model), an optimal operation problem was formulated to ensure adequate set-point (QD size) tracking, resulting in a narrow CSD. Overall, (a) the proposed multiscale model is in good agreement with the experimental results; (b) the CFD simulation highlights the region in which the SFC should be operated (*i.e.*, stable slug flow regime); and (c) the optimizer is capable of effective simultaneous set-point tracking (QD size) and disturbance rejection (S2S variation) while ensuring a narrow CSD.

In summary, the CFD model was used to identify a stable slug regime and quantify the S2S variation in a millifluidic SFC. Additionally, a dynamic multiscale model for a SFC was developed to describe the temporal evolution of average QD size and CSD. Lastly, an optimization-driven controller scheme was implemented to ensure that the QDs match the industry specifications while maintaining a narrow CSD. Overall, the proposed work is a major leap towards continuous nano-manufacturing of metal halide perovskite QDs, which will be crucial in meeting market demands of the optoelectronics industry in the coming decade.

Literature Cited:

1. Protesescu L, Yakunin S, Bodnarchuk MI, Krieg F, Caputo R, Hendon CH, Yang RX, Walsh A, Kovalenko MV. Nanocrystals of cesium lead halide perovskites ($CsPbX_3$, X= Cl, Br, and I):

novel optoelectronic materials showing bright emission with a wide color gamut. Nano Letters. 2015 Jun 10; 15(6):3692-6.
2. Pu Y, Cai F, Wang D, Wang JX, Chen JF. Colloidal synthesis of semiconductor quantum dots toward large-scale production: a review. Industrial & Engineering Chemistry Research. 2018 Feb 14; 57(6):1790-802.
3. Abdel-Latif K, Epps RW, Kerr CB, Papa CM, Castellano FN, Abolhasani M. Facile room-temperature anion exchange reactions of inorganic perovskite quantum dots enabled by a modular microfluidic platform. Advanced Functional Materials. 2019 Jun; 29(23):1900712.
4. Rasche ML, Jiang M, Braatz RD. Mathematical modeling and optimal design of multi-stage slug-flow crystallization. Computers & Chemical Engineering. 2016 Dec 5; 95:240-8.
5. Sitapure N, Epps R, Abolhasani M, Kwon JS. Multiscale modeling and optimal operation of millifluidic synthesis of perovskite quantum dots: towards size-controlled continuous manufacturing. Chemical Engineering Journal. 2020 Dec 7:127905.

632449 Exploring the Physical Aging Behavior of Hpmcas *Via* Thermal Analysis

Abstract:

A major challenge in the design and development of oral delivery drugs is the "solubility challenge," where a substantial amount of newly discovered small molecule drugs exhibits low solubility, resulting in low bioavailability. One strategy to increase the bioavailability of hydrophobic small molecule drugs is to form an amorphous solid dispersion (ASD). In this approach, amorphous active pharmaceutical ingredients (APIs) are typically dispersed into an amorphous polymer matrix. The glassy polymer matrix kinetically hinders the recrystallization of the API, keeping the API in its amorphous state. Amorphous API exhibits increased bioavailability and aqueous solubility in comparison to its crystalline counterpart. However, despite their kinetic stability, ASDs often remain thermodynamically unstable with a preferential phase separation into API-rich and polymer-rich domains under certain environmental conditions. We hypothesize that this unfavorable phenomenon is primarily facilitated by the molecular mobility of the glassy polymer as the system undergoes structural relaxation.

In this study, we examine this hypothesis by analyzing the physical aging behavior of one of the most common polymer excipients in ASDs, hypromellose acetate succinate (HPMCAS), using conventional differential scanning calorimetry and fast scanning calorimetry. Through temperature scans across the glass transition temperature (T_g) and isothermal aging below the T_g, we compute the aging rate and observe the enthalpy recovery behavior on heating. Contrary to theoretical expectations, when it is aged well below T_g, HPMCAS exhibits a unique devitrification mechanism. In this mechanism, the enthalpy lost on physical aging is recovered notably before T_g, indicating a recovery process different from the alpha relaxation. More importantly, this process is not attributed to a secondary relaxation process as it is not discrete but rather continuous such that at longer aging times, the devitrification shifts towards the alpha relaxation process. This unique behavior suggests a relaxation mechanism that may directly influence the stability of drug molecules within the HPMCAS structure and impact ASD design, processing, and storage.

625167 Accurate Temperature Prediction of the Freeze-Dried Cake during Secondary Drying Process in a Laboratory Scale Lyophilizer

Abstract:

Freeze drying is a widely used process that improves the stability of the pharmaceutical products. Conventional freeze-drying consists of two sequential steps: (1) primary drying where ice sublimates from the product to leave behind a porous cake (10-30 hr) and (2) secondary drying where the remaining bound water is removed from the porous cake (2-10 hr). Currently, there is a lack of robust models for secondary drying with comparable accuracy and flexibility as primary drying. To alleviate this issue, we investigated secondary drying using a lab-scale lyophilizer and developed an accurate heat transfer models of this process. Experiments involved lyophilizing various sugars (e.g., sucrose and mannitol) under different operating conditions to determine the effects they have on energy transfer during secondary drying. The overall heat transfer coefficient at the bottom of the vial was determined by using a heat flux sensor to measure the slope of heat flux versus temperature difference between the shelf and the cake. Experimental results show that the heat transfer coefficient is significantly different when compared to the primary drying process, and that the vial's thermal properties play a large role in the heat transfer. Lastly, we develop three theoretical models to describe the heat transfer in the vial based on solving the heat equation with bound water desorption, using published desorption kinetic parameters and equilibrium moisture content data. The models describe the temperature and moisture content in the lyophilized cake with varying degrees of complexity depending on the application of interest. Simulation results are validated by comparison of temperature predictions with experimental data for shelf temperatures $15^oC \leq T_{sh} \leq 35^oC$. These models have great potential for the precise estimation of the cake temperature and drying time compared to trial-and-error approaches. It is expected that our models will aid in understanding and optimizing secondary drying for the lyophilized product.

621795 Physicochemical Modeling of Drug Stability in Multilayer Polymeric Films Containing an Aqueous Moisture Barrier Layer

Abstract:

Designing a drug dosage form requires the inclusion of excipients that provide enhanced drug solubilization, drug stabilization, controlled release, flavor masking, dilution, among many other properties. The inclusion of these additives can often result in complex interactions requiring significant formulation development to identify the correct combination of additives for a specific active pharmaceutical ingredient. An alternative to this approach is to prepare excipients into discrete layers that provide their relative benefits without needing to identify all blend compatibilization and concerns. Multilayer films provide advantage of incorporating separate additives in separate layers in order to provide different functions such as enhancing aqueous solubility and bioavailability of Biopharmaceutical Classification System (BCS) Class-II and Class-IV drugs in the form of soluble polymeric drug dispersions, and including a beneficial layer of moisture barrier to maintain drug stability in storage. This separation also provides ease of developing mathematical modeling considering only one-dimensional transport for predictive impact during industrial formulation ensuring specific drug release profiles and stability requirements. Here, we model and develop a barrier layer to prevent/minimize water vapor permeation to the active layer so that water sensitive APIs can be protected to a stipulated level during the long-term storage.

Multilayer drug loaded films consisting polyvinyl alcohol (PVA) active layer and tannic acid modified chitosan barrier layer (both layers were prepared by solvent casting) are fabricated using a uniaxial heat press. Active and barrier layers are laminated by putting edible starch-based hot-melt adhesive made from powdered corn-syrup and fructose. Water vapor transmission through the barrier layer was determined by ASTM E96M moisture transmission test at a relative humidity of 70%. Water vapor transmission through the barrier layers is modeled by unsteady-state Fickian diffusion equation coupled with mass transfer at the surface and interface between two layers, and drug impurity formation kinetics. The partial differential equation (PDE) is solved by the method of lines, where ordinary differential equations (ODEs) obtained after spatial discretization are coupled with the ODEs from drug impurity formation kinetics. The water vapor concentration profiles in the barrier layer are used to find mass flux, mass flow rate and cumulative water transmission to the active layer. Data from the experimental water uptake by the active layer and impurity formation were fitted to the model profiles to optimize the transport (diffusivity), mass transfer (surface mass transfer coefficient, partition coefficient) and reaction kinetic constants based on <1% impurity formation in a year. Finally, relationship between film properties (types, thickness, moisture absorption, delamination upon solvation) and user provided requirements were established in order to develop a generalized procedure for drug formulations.

626788 Development of an HME Based Extended-Release Formulation Using Experimental and in silico Approach

Abstract:

Introduction

Hot melt extrusion (HME) is a continuous manufacturing process primarily carried out using co-rotating intermeshing twin-screw extruders (TSE). In the pharmaceutical context, HME is primarily used to produce amorphous solid dispersions of poorly soluble active pharmaceutical ingredients (APIs). In addition, it can also facilitate the development and production of products with crystalline API embedded into a polymer matrix and nanopharmaceuticals. Being a continuous manufacturing technology, HME-based drug development typically requires prohibitively high amounts of API (kilograms of premix vs grams available in the early development stages) for successful formulation screening, early process development and scale-up. The process setup itself is modular, allowing the process to be tailor-made for any formulation. This might however be problematic in the context of choosing the appropriate and efficient setup for an unknown formulation and it is an additional reason for relatively high quantities of material usually needed for the screening. Effectively solving the multidisciplinary challenge of formulation development, early process screening, effective scale-up and transfer to GMP production represents one the key challenges of the pharmaceutical industry and is especially challenging for novel and imported production technologies like HME.

Methodology

Addressing these challenges, our groups have worked on the development of *in silico* and experimental tools for simpler process development and scale-up. Since to this date most of the process setup and scale-up activities are performed experimentally and empirically, one of the goals was to create *in silico* tools for a rational, science-based process setup and scale-up, while addressing other important aspects, such as API degradation and overall product quality [1]–[3]. The fundamental idea behind process modelling is the breakdown and the detailed analysis of the key process aspects, among which the most prominent might be the analysis of flow patterns developed as a result of the rotation and geometry of the individual screw element pairs [1], [4]–[9]. In this work we′ve followed the development of a novel extended-release product for oral administration. The process development starts with extrusions on the 9mm tabletop extruder (Three Tec ZE9) and continues to 16mm (Leistritz NANO16) and 18mm (Coperion ZSK18) extruders. The experimental development is closely followed with 3D Smoothed Particle Hydrodynamics (SPH) simulations of individual screw elements that comprises the screw configuration, and reduced-order 1D HME process simulations.

Results

The experimental setup on all three extruders is investigated in detail using information collected during the experiments and simulations. The produced extrudates are investigated in term of the material science, release profile, and stability. The extrudate quality is then connected to the prevailing extrusion state with the help of reduced order 1D HME simulations. The process was also scaled from the tabletop 9mm extruder to a triple-flighted 16mm extruder and a 18mm pilot plant size extruder. The differences found in the performance of the individual screw elements, differences in process setup and extruder design and scale are investigated and connected to the resulting extrudate quality.

Literature

[1] J. Matić, A. Witschnigg, M. Zagler, S. Eder, and J. Khinast, "A novel in silico scale-up approach for hot melt extrusion processes," Chem. Eng. Sci., vol. 204, pp. 257–269, Aug. 2019.

[2] J. Matić, A. Paudel, H. Bauer, R. A. L. Garcia, K. Biedrzycka, and J. G. Khinast, "Developing HME-Based Drug Products Using Emerging Science: a Fast-Track Roadmap from Concept to Clinical Batch," AAPS PharmSciTech, vol. 21, no. 5, p. 176, Jul. 2020.

[3] J. Matić et al., "Towards predicting the product quality in hot-melt extrusion: Small scale extrusion," Int. J. Pharm. X, vol. 2, p. 100062, Dec. 2020.

[4] J. Matić *et al.*, "Towards predicting the product quality in hot-melt extrusion: Pilot plant scale extrusion," *Submitted.*

[5] A. Eitzlmayr and J. G. Khinast, "Co-rotating twin-screw extruders: Detailed analysis of conveying elements based on smoothed particle hydrodynamics. Part 1: Hydrodynamics," Chem. Eng. Sci., vol. 134, pp. 861–879, Sep. 2015.

[6] A. Eitzlmayr and J. G. Khinast, "Co-rotating twin-screw extruders: Detailed analysis of conveying elements based on smoothed particle hydrodynamics. Part 1: Hydrodynamics," Chem. Eng. Sci., vol. 134, pp. 861–879, Sep. 2015.

[7] A. Eitzlmayr, J. Matić, and J. G. Khinast, "Analysis of flow and mixing in screw elements of corotating twin-screw extruders via SPH," AIChE J., vol. 63, no. 6, pp. 2451–2463, Jun. 2017.

[8] A. Eitzlmayr et al., "Experimental characterization and modeling of twin-screw extruder elements for pharmaceutical hot melt extrusion," AIChE J., vol. 59, no. 11, pp. 4440–4450, Nov. 2013.

[9] A. Eitzlmayr et al., "Mechanistic modeling of modular co-rotating twin-screw extruders," Int. J. Pharm., vol. 474, no. 1–2, pp. 157–176, Oct. 2014.

[10] R. Baumgartner, J. Matić, S. Schrank, S. Laske, J. Khinast, and E. Roblegg, "NANEX: Process design and optimization," Int. J. Pharm., vol. 506, no. 1–2, pp. 35–45, Jun. 2016.

628373 Multi-Objective Supply Chain Optimization in Personalized Healthcare

Abstract:

Autologous Chimeric Antigen Receptor T cell therapies are a type of immunotherapy, where T cells are taken from a patient's blood and engineered to express the Chimeric Antigen Receptor (CAR). CAR T cells are then capable of targeting and killing cancer cells [1]. The two first commercial products have been approved in 2017 and 2018 by the U.S. Food and Drug Administration (FDA) and he European Medicines Agency (EMA) respectively for the treatment of B-cell lymphoma [2,3]. More recently, the U.S. FDA approved two more treatments for relapsed or refractory mantle cell lymphoma (2020) and for relapsed or refractory large B-cell lymphoma (2021) [4].

Unlike batch produced pharmaceuticals, CAR T cell therapies use patient cells as the starting material, placing the patient schedule in the center of the decision-making process. This unique 1-to-1 production and supply chain model translates into manufacturing lines and distribution nodes designed for and occupied by a single therapy, posing hurdles to the development of robust and responsive networks. Moreover, the predicted increase of patient numbers, the sensitive nature and short shelf-life of these therapeutics pose additional challenges to the coordination of manufacturing, storage and distribution [5].

In this work, we present a Mixed Integer Linear Programming (MILP) model for the optimal design of a supply chain network of CAR T cell therapies under different demand scenarios. The problem is formulated as multi-objective optimization aiming to minimize the average therapy cost (ATC) and the average return time of the therapies (ART). The preliminary results showcased in Figure 1 highlight the trade-off between the two objectives. The results are reported for two annual demand profiles of 200 (triangles) and 500 (squares) patients and two different manufacturing durations (TMFE). The base case scenario represents the current state of the art of 13 days of manufacturing (filled symbols), while a forward-looking scenario where the manufacturing time is reduced to 7 days is also considered (empty symbols).

We can observe that in the base case scenario an increase in the demand leads to an increase of the average therapy cost for the same ATR, which indicates that for prolonged manufacturing time the effect of economies of scale is not significant. On the other hand, in the forward-looking scenario it is observed that for long manufacturing times an increased demand may lead to facilities with capacity in order to satisfy all patients in time.

Future directions include: (i) investigation of scenarios with higher number of patients per year; and (ii) assessment of the effect of patient-specific manufacturing time on the supply chain network. The second point is particularly important, since the patient conditions at the time the cell sample is taken influence significantly cell doubling time and consequently the total manufacturing time required. The results will help identify bottlenecks in the overall manufacturing and distribution process, as well as the impact of a variable manufacturing time on the overall network. Finally, each scenario is simulated for different demand profiles in order to quantify the sensitivity of the patients scheduling for a given total demand and manufacturing time.

Acknowledgments

Funding from the UK Engineering & Physical Sciences Research Council (EPSRC) for the Future Targeted Healthcare Manufacturing Hub hosted at University College London with UK university partners is gratefully acknowledged (Grant Reference: EP/P006485/1). Financial and in-kind support from the consortium of industrial users and sector organizations is also acknowledged.

References

[1] B. L. Levine, J. Miskin, K. Wonnacott, and C. Keir, "Global Manufacturing of CAR T Cell Therapy," Mol. Ther. - Methods Clin. Dev., vol. 4, no. March, pp. 92–101, 2017.

[2] Novartis, "KYMRIAH Treatment Process, Dosing & Administration | HCP," 2018. [Online]. Available: https://www.hcp.novartis.com/products/kymriah/acute-lymphoblastic-leukemia-children/dosing-and-administration/. [Accessed: 24-Mar-2021].

[3] Kite Pharma, "First CAR T Therapy for Certain Types of Relapsed or Refractory B-Cell Lymphoma," 2018. [Online]. Available: https://www.yescartahcp.com/. [Accessed: 24-Mar-2021].

[4] A. Mullard, "FDA approves fourth CAR-T cell therapy". Reviews Drug Discovery 20, 166 (2021)

[5] R. Griffiths and M. Lakelin, "Successfully managing the unique demands of cell therapy supply chains." p. 9, 2017.

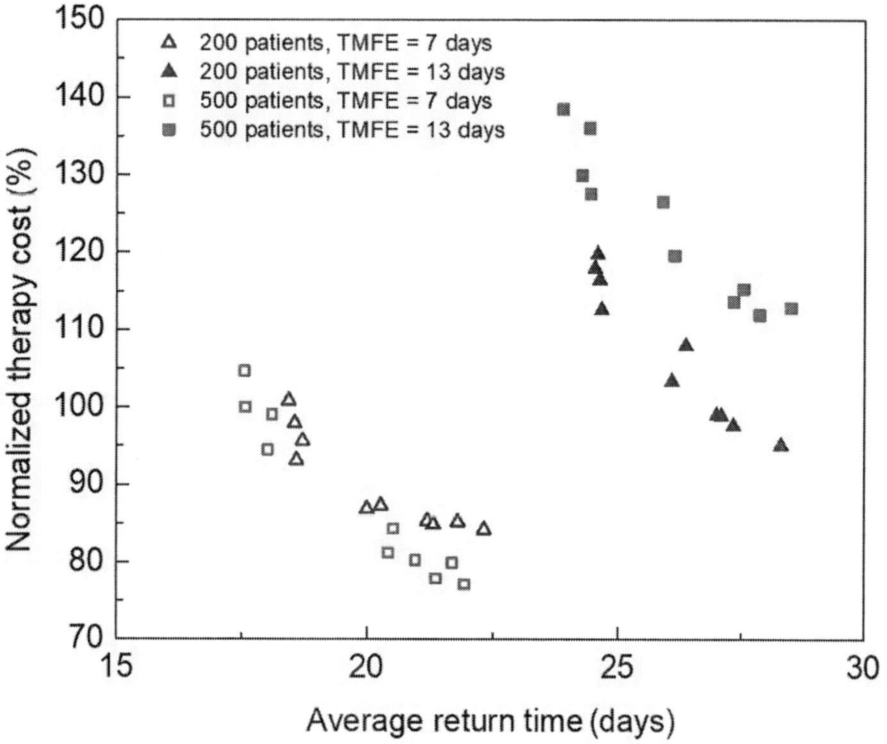

Figure 1: normalized therapy cost versus average return time of a therapy for two demand scenarios and two total manufacturing time excluding quality control.

630367 Robust Bioprocessing of Lignocellulose Using Microbial Tipping Points

Abstract:

Biofuels derived from lignocellulosic feedstocks have been shown to have near zero and in some cases net negative greenhouse gas emissions, driving public and private interest in producing lignocellulosic biofuels on an industrial scale. Anaerobic digestion (AD) is an established and scalable biological process for biogas production from a variety of feedstocks, including lignocellulosic biomass. AD uses a mixed microbial community adapted to the feedstock and operating conditions to produce biogas in four basic steps, hydrolysis, acidogenesis, acetogenesis, and methanogenesis. Biogas is on average 60 vol% methane and 40 vol% carbon dioxide and can be upgraded to renewable natural gas (RNG) by downstream carbon dioxide removal.

Lignocellulosic feedstocks are recalcitrant to hydrolysis, which is often the rate-limiting step for AD of this material. For other feedstocks, e.g. wastewater sludge, food waste, or manure, the rate-limiting step is considered to be methanogenesis. Because of this difference, the process conditions required for microbiomes to achieve high methane production rates during AD of lignocellulosic feedstock may have unique pH, temperature, residence time, and loading rate optima and tolerances. Hydrolysis-limited microbiomes may also have different sensitivities to perturbations of these conditions compared to more prevalent wastewater and manure systems. Understanding the sensitivities of AD to upsets or tipping points is necessary for operating a robust process, with robust defined as being able maintain the key process performance parameters without easily upsetting beyond recovery. A tipping point is defined as the process parameter value at which a biological community becomes unstable in terms of its key product(s) formation. This study proposes a method to define the tipping points for an AD microbiome fed senescent switchgrass. Within those tipping points the variation in process variables avoids irreversible reductions in system performance.

The environmental stresses evaluated in this study are temperature, pH, and overfeeding. These stresses were chosen as those likely to occur during full-scale operation, and that have been known to upset conventional waste digesters. Tipping points will be defined at similar thresholds as those used in inhibitor studies, with tipping points at 50% and 90% decline in methane production after the perturbation is applied. The use of two levels of failure is important as some stresses may not cause catastrophic failure within an operationally relevant range of stress magnitudes. The volatile fatty acid (VFA) concentration will also be monitored as a secondary variable used to identify and characterize a failing digester, as VFAs represent the key intermediates between the potential rate-limiting steps of hydrolysis or methanogenesis. For the pH and temperature tipping point experiments, both positive and negative changes will be evaluated, while for overfeeding only positive changes will be explored. The magnitude of each shock is provided in Table 1. Based on prior AD literature, lower feeding rates are unlikely to result in the dramatic and catastrophic failures that higher feeding rates and both higher and lower pH and temperature conditions can.

Each shock condition will be carried out in duplicate 1-L semi-continuous bioreactors and compared to duplicate controls. The digestate from a pilot scale anaerobic digester fed switchgrass at 55°C, 8.5 pH, and a 10-day retention time will be used to inoculate the 1-L bioreactors. After the shock is administered for a set time the bioreactors will be returned to the process conditions of the pilot reactor and monitored for three retention times for process recovery. Biogas production will be continuously monitored while the VFA concentration, feedstock conversion, and biogas composition are measured at 0.01, 0.1, 0.5, 1, 2, and 3 retention times.

This work aims to define the range of perturbation that a lignocellulosic AD system can experience without significant and irreversible loss of methane production. The method described allows for the sensitivity of a microbial system to be defined in terms of control variables. This then informs how precisely the system must be controlled to avoid biological tipping points leading to loss of product. Defining these tipping points and controlling within them will increase the operating efficiency of microbial systems and enhance the technical and economic feasibility of converting recalcitrant lignocellulosic feedstocks to biofuels.

Table 1. The process variable settings for each tipping point experiment

Process Variable	Settings
Overfeeding (g VS/L)	+10, +20, +30
Temperature (°C)	±10, ±20, ±30
pH	±0.5, ±1.0, ±1.5

624307 Keynote: Multi-Objective Optimization, State Estimation, and Advanced Control of a Semi-Batch Process for the Enzymatic Conversion of Lactose into Value-Added Products

Abstract:

Some of the main challenges experienced when scaling a laboratory enzymatic process to commercial scale include achieving optimal operations, performing online monitoring, and ensuring process controllability. Often times, laboratory-scale enzymatic processes correspond to simply proof of concept experiments that are carried out to characterize the reaction phenomena and develop preliminary process models. For example, in laboratory settings, ideal process conditions, such as negligible spatial variation in concentration, temperature or pH are much more achievable than at industrial scales[1]. As a result, when applying these models to industrial scales, a higher level of uncertainty and error is expected when compared to laboratory scale. Additionally, when developing these models, highly specific offline analytical methods are employed and relied upon due to their increased precision and accuracy over traditional online monitoring methods. However, in many industrial processes, soft sensors or estimation methods must be used online to infer process states from measurements of variables such as the oxygen uptake rate or carbon dioxide production rate. When using these methods, an additional level of uncertainty is introduced due to reliance on calibration ranges and model accuracy[2]. Thus, considering all these factors, it is often challenging to predict if laboratory scale processes can be feasibility scaled up. In this work, a process optimization, state estimation, and advanced control framework is introduced to determine the feasible scaled-up operations of a laboratory enzymatic process, considering a new enzymatic model for the conversion of waste lactose into a series of value-added products known as galactooligosaccharides (GOS).

The enzymatic model used in this work utilizes β-galactosidase to convert lactose into GOS products while distinguishing purely galactose derived di-, tri-, and tetrasaccharides from those containing a glucose group[3]. Due to the structural similarities of these isomers, online monitoring techniques are unable to isolate and uniquely distinguish the specific isomer concentrations. Instead, such techniques are only able to measure the combined concentrations of the isomers and provide one total concentration, thus posing a more challenging monitoring problem. Additionally, this specific model was chosen as it is currently at laboratory scale and could potentially be scaled to a commercial level to address the lactose waste problem plaguing small-scale cheese manufacturers[4].

Due to the many process uncertainties, such as raw feed cost and enzyme cost in this preliminary stage, conducting a robust and accurate economic analysis is challenging and would likely exhibit a wide distribution of economics, ultimately providing unreliable results. Instead, to examine the feasibility of scaling up this process, a multi-objective optimization is carried out considering 2 main factors: maximizing lactose conversion and GOS production rate as these will have the greatest effect on the process economics. Using the results of this multi-objective optimization, a trajectory-based model predictive controller (MPC) is implemented to guarantee that the process adheres to the optimum profile and is able to respond to simulated disturbances that are typical of industrial processes. Finally, a Moving Horizon Estimator (MHE)[5] is implemented to cope with the noise from the online measuring techniques and to decouple the combined isomer concentrations, providing insight into the specific compound concentrations in the reactor for optimal operations and control.

References

[1] Villadsen, J., Nielsen, J., Lidén, G. (2011). *Bioreaction Engineering Principles, 3.* Springer, Boston, MA.

[2] Stanke, M., Hitzmann, B. (2013). Automatic Control of Bioprocesses. In: Measurement, Monitoring, Modelling and Control of Bioprocesses, *Advances in Biochemical Engineering/Biotechnology, 1.* Springer, Berlin, Heidelberg.

[3] Schultz, G., Alexander, R., Lima, F. V., Giordano, R. C., Ribeiro, M. P. A. (2021). Kinetic Modeling of the Enzymatic Synthesis of Galacto-Oligosaccharides: Describing Galactobiose Formation. *Food and Bioproducts Processing, 127*, 1-13.

[4] Illanes, A. (2011). Whey Upgrading by Enzyme Biocatalysis. *Electronic Journal of Biotechnology. 14.*

[5] Alexander, R., Campani, G., Dinh, S., Lima, F.V. (2020). Challenges and Opportunities on Nonlinear State Estimation of Chemical and Biochemical Processes. *Processes, 8, 1462.*

631677 Keynote: Virtual-Engineering Software Framework for Integrated Biomass Conversion Modeling

Abstract:

This presentation covers the design and implementation of a software tool to systematically connect computational models of unit operations to simulate an integrated process of low-temperature conversion of biomass to fuel. This virtual engineering (VE) software was designed with the overarching goal of connecting unit models written in various programming languages and requiring different computational resources within a single, flexible framework. The models and features currently considered for the VE library include mechanistic models for pretreatment, enzymatic hydrolysis, and aerobic bioreaction; high-fidelity computational fluid dynamics (CFD) simulations for enzymatic hydrolysis and aerobic bioreaction; and the capability to perform techno-economic analyses (TEA) using Aspen Plus, a commercial software package. The CFD models require access to high-performance computing (HPC) resources, so in addition to handling multiple programming languages and interfaces, the VE software must also be capable of interacting with an HPC scheduler to submit, run, and post-process jobs.

Using the Python programming language, a new VE software package has been developed that contains functionality to manage the input-output communication between various unit models, schedule simulations to run on NREL's HPC and analyze those results, and interface with existing TEA software workflows. A Jupyter-notebook GUI was also created to solicit user input and provide documentation. In cases where multiple models for a particular unit-operation exist, selection between models is accomplished through a simple checkbox, with the appropriate inputs and outputs being parsed and converted seamlessly in the background. Each operation makes use of a different programming language, but the flow of information from pretreatment to enzymatic hydrolysis to bioreaction is managed with an intuitive, centralized file-communication strategy. In this talk, the programming approach and implementation details of the notebook are presented for multiple possibilities of the conversion process, including a demonstration of the ability to manage HPC resources. Additionally, an example of a sensitivity study of treatment parameters governing the overall conversion outcome is shown which highlights the ease of defining new problems using the VE Notebook workflow and leads into a discussion of ongoing work to enable outer-loop optimization studies.

631135 Optimizing Sampling Efficiency in the Einstein Crystal Free Energy Method

Abstract:

The Einstein Crystal Method (ECM) is an atomistic simulation approach for computing the free energy of molecular solids that has received significant interest in recent years. The ECM has been used in the last decade to compute the melting temperature of ice, assess relative polymorph stabilities at ambient temperature, and estimate the absolute solubility of pharmaceutical compounds in solution among other properties. ECM calculations can be quite computationally expensive, and therefore a key to successfully using the ECM for property estimation is achieving efficient sampling along the entire thermodynamic path from the physical crystal to the ideal reference state. In particular, the efficiency of the sampling in the ECM is extremely sensitive to the lambda scheduling function used for the various alchemical steps in the method.

In this work, we assess a variety of different lambda scheduling functions for the various steps in the ECM including adding all-atom harmonic restraints as well as decoupling the various interactions in the system. We present the overlap matrices corresponding to different scheduling functions for each step to show the most efficient lambda spacing. Finally, we present an iterative lambda schedule optimization approach that can adaptively adjust lambda scheduling from a short simulation to optimize the overlap across adjacent lambda windows for a subsequent longer production run. These improved lambda scheduling functions have the potential to significantly improve the efficiency of all future ECM calculations.

623046 Physics-Based Computational Models to Expedite Pharmaceutical Solid-Form Selection

Abstract:

Improving poor physicochemical properties of active ingredients and intermediates is a challenging problem for solid formulation scientists. Physics-based computational models to select pure solvents, solvent mixtures and coformers for crystallization processes has recently attracted much attention in the pharmaceutical industry. Here, I will present some case studies to demonstrate how we employ physics-based models to (i) perform a rational solvent selection to identify solvents with high and low probability to form solvates and/or impurity purge; and (ii) select coformers to form a cocrystal and guide comprehensive experimental solid form screening to mitigate challenges in solid formulation.

624574 Development of RTD-Based Flowsheet Modeling Including Process Uncertainty for Continuous Solid-Based Drug Manufacturing

Abstract:

Over the past twenty years, there has been a trend in pharmaceutical industry to commercialize continuous manufacturing (CM) for the production of solid-based drug products, given the various advantages of CM and the alignment with FDA's support of the quality-by-design paradigm (Lee et al., 2015; Plumb, 2005; Schaber et al., 2011). Overall, CM has the potential to reduce production costs while improving product quality (Ierapetritou, Muzzio, & Reklaitis, 2016; Schaber et al., 2011). To fully utilize the above benefits for CM, it is important to properly design the process to guarantee the production of qualified drug products. Process design relies heavily on detailed understanding of the process characteristics, which gives rise to the need for developing process models. A variety of modeling techniques have been developed in the literature, including discrete-element method (DEM), population balance modeling (PBM), residence time distribution (RTD) modeling, and semi-empirical modeling (Ierapetritou et al., 2016). We focus on RTD modeling in this study because it is well-suited for building integrated flowsheet models with low computational cost while maintaining sufficient degree of characterization of the process behavior, particularly from a process dynamics point of view. This implies that given an upstream disturbance, the system response can be predicted using convolution of the developed RTD models with the disturbance. A number of studies have been conducted to develop RTD-based flowsheet models for different manufacturing routes (Martinetz et al., 2018; G. Tian et al., 2021; Geng Tian et al., 2017), and disturbance propagation has been tracked downstream to quantify its effects on final drug product quality.

RTD model parameters are determined based on RTD experiments conducted under steady-state operation. Maintaining a steady state during manufacturing can get difficult due to numerous unpredictable events like process fluctuation of flowrates, changes in flow patterns within unit operations and human-based variabilities. Thus, there is a certain degree of uncertainty inherently associated with the system and corresponding RTD models. The uncertainty in RTD characterization can lead to significant implications for RTD flowsheet applications like material traceability and diversion of out-of-specification (OOS) material. The uncertainty in RTD of process flowsheets can prohibit identifying the true start and endpoints of OOS material, leading to production of OOS tablets. Thus, it is important to consider this inherent process uncertainty and properly characterize its range and implications on drug product quality. To this end, we have developed a two-stage methodology encompassing quantification of the degree of uncertainty associated with RTD profiles for various unit operations and investigation of uncertainty propagation along the downstream unit operations of the manufacturing line to obtain the overall RTD flowsheet model incorporating process uncertainty. The updated RTD flowsheet model can then be used to demonstrate its ability for disturbance propagation and precise determination of OOS products. The proposed methodology would pave the way for robust and efficient solid-based drug production as it provides a unique strategy to incorporate the effects of process uncertainty in maintaining drug product quality for CM applications.

References:

Ierapetritou, M., Muzzio, F., & Reklaitis, G. (2016). Perspectives on the continuous manufacturing of powder-based pharmaceutical processes. *AIChE Journal, 62*(6), 1846-1862. doi:10.1002/aic.15210

Lee, S. L., O'Connor, T. F., Yang, X., Cruz, C. N., Chatterjee, S., Madurawe, R. D., . . . Woodcock, J. (2015). Modernizing Pharmaceutical Manufacturing: from Batch to Continuous Production. *Journal of Pharmaceutical Innovation, 10*(3), 191-199. doi:10.1007/s12247-015-9215-8

Martinetz, M. C., Karttunen, A. P., Sacher, S., Wahl, P., Ketolainen, J., Khinast, J. G., & Korhonen, O. (2018). RTD-based material tracking in a fully-continuous dry granulation tableting line. *Int J Pharm, 547*(1-2), 469-479. doi:10.1016/j.ijpharm.2018.06.011

Plumb, K. (2005). Continuous Processing in the Pharmaceutical Industry. *Chemical Engineering Research and Design, 83*(6), 730-738. doi:10.1205/cherd.04359

Schaber, S. D., Gerogiorgis, D. I., Ramachandran, R., Evans, J. M. B., Barton, P. I., & Trout, B. L. (2011). Economic Analysis of Integrated Continuous and Batch Pharmaceutical Manufacturing: A Case Study. *Industrial & Engineering Chemistry Research, 50*(17), 10083-10092. doi:10.1021/ie2006752

Tian, G., Koolivand, A., Gu, Z., Orella, M., Shaw, R., & O'Connor, T. F. (2021). Development of an RTD-Based Flowsheet Modeling Framework for the Assessment of In-Process Control Strategies. *AAPS PharmSciTech, 22*(1), 25. doi:10.1208/s12249-020-01913-8

Tian, G., Lee, S. L., Yang, X., Hong, M. S., Gu, Z., Li, S., . . . O'Connor, T. F. (2017). A dimensionless analysis of residence time distributions for continuous powder mixing. *Powder Technology, 315*, 332-338. doi:10.1016/j.powtec.2017.04.007

624616 Computational Fluid Dynamics Study to Resolve Mixing and Scale-up Challenges of Non-Newtonian Fluids in Pharmaceutical Industry

Abstract:

The generic pharmaceutical industry relies on the speed and rigor in the drug development process. Getting it right the first time requires detailed product and process understanding. Various unit operations like sterilization, milling, mixing, etc. govern the rapidity of the process. Mixing is one such operation, developing a proper technical understanding can yield highly productive results and can reduce experimental work thereby proving to be cost and time effective. This work provides a comprehensive approach for understanding mixing in different equipment for several parameters and scale-up strategy using computational fluid dynamics.

The first question to ask is what comprises good mixing. The judging parameters include the uniform distribution of energy with less dissipation and getting the right CQAs in the samples. The major challenges in mixing can be attributed to non-uniformity and unpredictability in scale-up operations. Computational fluid dynamics (CFD) is an approach that is coming into action in the pharmaceutical industry. CFD Software (ANSYS in this example) can be leveraged to understand mixing profiles and help in analyzing the technical aspects and provide a generalized methodology for the problem.

As the complexity of molecules has increased and with process intensification, the need to develop technical solutions faster and having a standardized approach will support the entire industry to reach its goal of serving people better. This study gives a comparison of equipment based on histograms, contour plots, and vector plots for shear stress, velocity, and molecular viscosity, along with the discrete phase model to draw inferences and conclusions for non-Newtonian fluid systems. Transient simulations are used to determine the particle distribution in the mixing vessel volume being dispersed from a specified location. This helps in finding time to achieve uniformity throughout the volume. Using CFD the problem at lab scale is resolved by determining the CPPs which in turn gives us the ability to scale-up the product in a plant to achieve desired targets.

The impact of this study can be huge, as it can contribute not only to reduce experimental work at lab scale but can also improve visibility in the scale-up plant operations. It further adds as a standardized method for such issues.

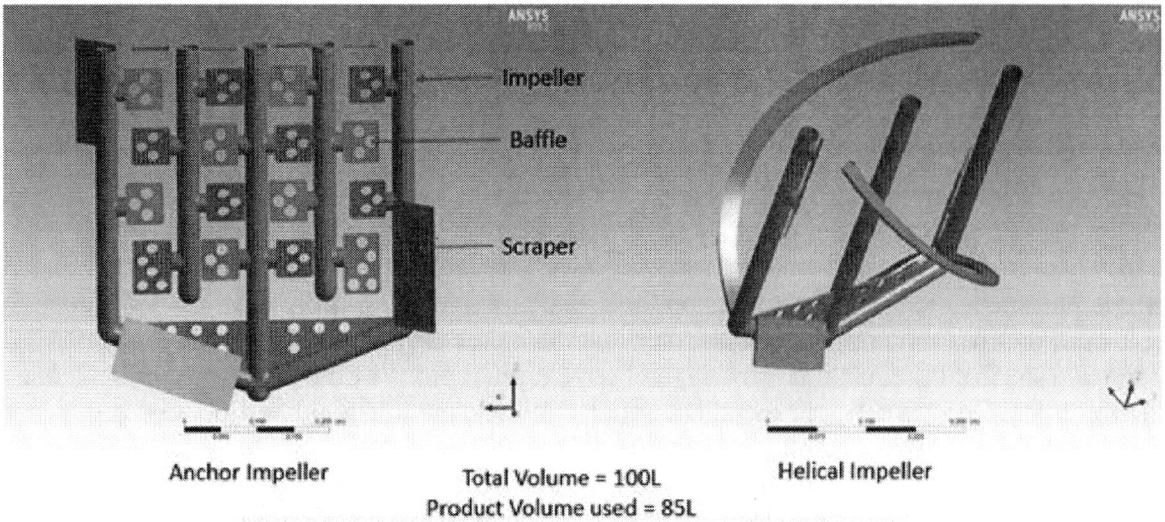

Figure 1: CAD models of Anchor and Helical impellers available at plant scale

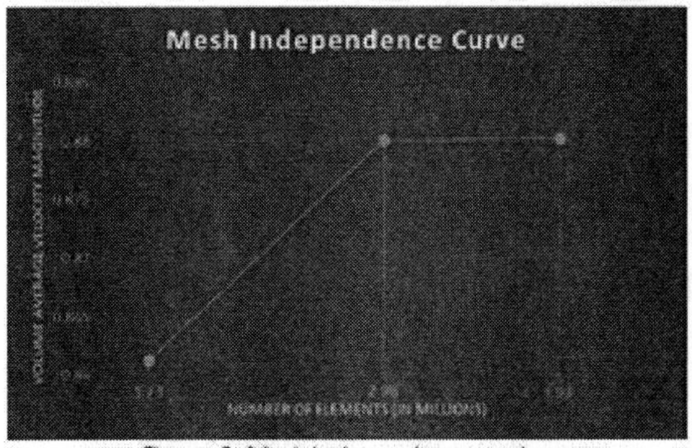

Figure 2: Mesh independence study

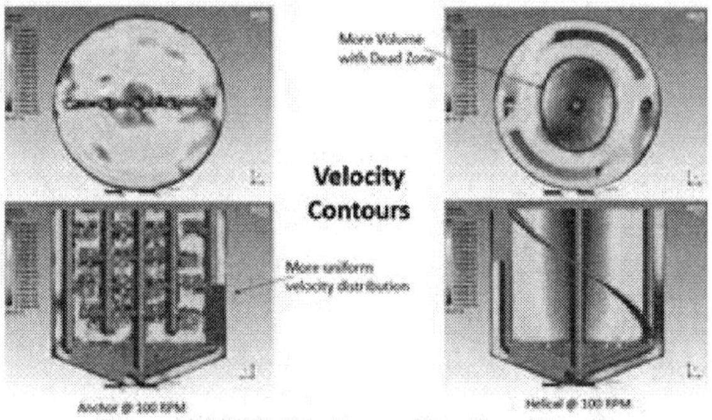

Figure 3: Velocity Contours of plant scale study

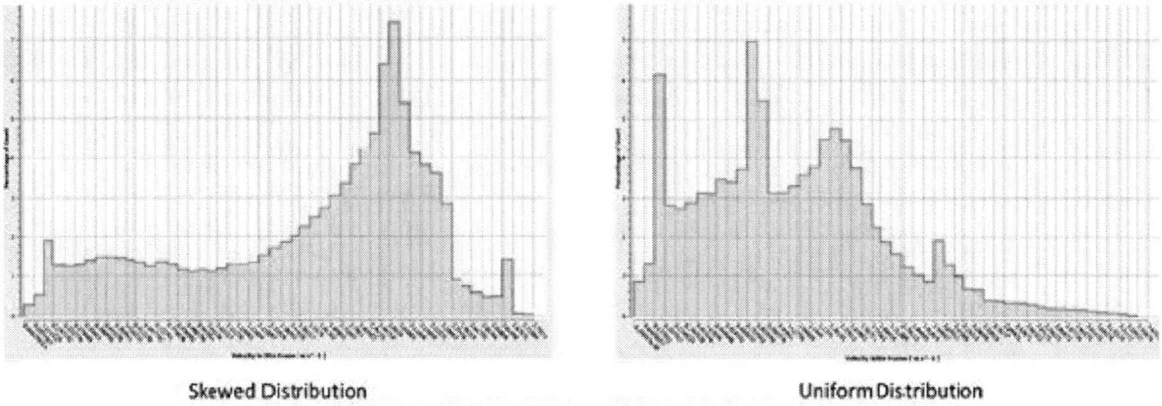

Figure 4: Velocity Histograms for Helical and Anchor impellers for plant scale study

631471 Manufacturing and Design of a COVID-19 Vaccine Antigen for Global Access

Abstract:

To address the COVID-19 pandemic globally, the world will require >10 billion doses of vaccines for SARS-CoV-2. This requirement exceeds the annual capacity for manufacturing vaccines for all other diseases by >2x. Most of these doses are required in low- and middle-income countries (LMICs), where platforms for the manufacturing and distribution of vaccines are limited. To fulfill this need, it is clear that further innovations in vaccine design and manufacturing are required. Protein subunit vaccines are promising candidates for global distribution because they can be manufactured at low costs in microbial hosts. The yeast *Komagataella phaffii* (*Pichia pastoris*), for example, is commonly used to manufacture protein therapeutics at large scales (~10^9 annual doses) in LMICs. We used *K. phaffii* to produce the SARS-CoV-2 spike protein receptor binding domain (RBD), a promising vaccine antigen found in many recent vaccine designs because of its small size and high density of neutralizing epitopes.

Here, we present a two-pronged approach of host-cell and protein engineering to improve global access to RBD-based vaccines. We genetically engineered *K. phaffii* to secrete the RBD without typical methanol-induction, which improved upstream titers ~5x and enabled longer production campaigns. We also rationally engineered an RBD protein with improved manufacturing titers and enhanced immunogenicity and antigenicity, enabling smaller dose sizes for wider vaccine distribution. The platform for RBD manufacturing presented here has been transferred to a GMP facility for manufacturing and clinical trials of engineering RBD antigens. This translational success suggests that the strain and protein engineering techniques presented here will be applicable to a wide range of protein therapeutics.

632519 Model Assisted Development for Downstream Processing of Therapeutic Proteins

Abstract:

In the last decade, many model-based approaches describing the development, robustness or other aspects of bioprocess unit operations have been developed, applied, and presented. These tools, relying on statistical or mechanistic principles, tend to concentrate on certain facets of single steps, and do not necessarily represent the reality of the multi-step bioprocess. Over the last year, we have focused on combining many of these models to enable process development for the entirety of the downstream process.

In this talk, we will discuss the creation and application of a new model assisted development strategy for the downstream process. By understanding the chemical properties of the different buffer solutions, as well as the interactions between proteins and chromatographic surfaces, we can predict, and with minimal experimental work, confirm the performance of protein A and ion exchange chromatography steps. These tools are presented along with design and validation experiments. In the second part of our talk, we describe the combination of these models into a single entity. This allows us to fully reproduce and optimize the multi-step downstream process as a whole and in silico. The robustness and sensitivity to input changes are then analyzed in terms of product quality, productivities, and process costs.

633051 Enhanced Process Characterization through Data-Rich Experimentation and Process Modeling: Generating an Intermediate Towards Islatravir

Abstract:

Tight development timelines and increasing expectations around process knowledge at time of filing have driven investments in data-rich experimentation tools aimed to maximize knowledge capture from each experiment. Commonly reaction characterization is accomplished through design of experiment methodologies coupled with statistical analyses and response surface modeling. Typically, however, these statistical analyses are restricted to single timepoints. Recently, automated reactor and sampling technologies have enabled scientists to capture temporal data throughout each experiment. This greater data density has enabled the generation of data-driven models through techniques such as dynamic surface response methodology (DRSM) and the leveraging of these models to improve process robustness and efficiency [[1]].

Herein, we present the use of reaction and sampling automation to generate an extensive reaction dataset during the completion of a design of experiments as part of process characterization of one of the penultimate intermediates in the synthesis of islatravir, a late-stage drug substance candidate. Data-driven models generated from this dataset were used to inform reaction stability and drive process decisions.

[[1]] J. Jurica, J. McMullen. 2021. *Org. Proc. Res. Dev.,* 25, 2, 282-291

631484 Impurity Tracking in Integrated Pharmaceutical Batch Processes Subject to Raw Material Variability

Abstract:

In recent years, mechanistic models have been applied to a broad range of applications in pharmaceutical R&D and engineering, improving efficiency through reduced data requirements, facilitating scale-up and tech transfer, and enabling virtual design space exploration. Many pharmaceutical applications of mechanistic models focus on a single unit operation, using a component model for activities such as scale-up. However, in recent years, integrated or system models have been used to describe a series of unit operations or even the end-to-end process. These integrated models allow for a holistic assessment of the effects of upstream unit operations on product attributes and allow a combination of process parameters across multiple steps to be explored. Such system models are typically associated with continuous processes and are constructed through flowsheet modeling. However, using these same methodologies, integrated flowsheet models of batch processes can be constructed, specifying the schedule of each batch step.

In this talk, an integrated system model of a batch pharmaceutical manufacturing process will be presented, including both the drug substance and drug product manufacturing steps. This system model will be used to demonstrate impurity formation in a synthesis step, possible rejection through purification steps, tracking of the impurity through subsequent processing steps and ultimately the presence of the impurity in the drug product. Further, a sensitivity analysis is performed to identify critical process parameters (CPPs) on the critical quality attribute (CQA) of the impurity's presence in the final product. Visualizations of the design space are presented and analyzed to identify risk mitigation strategies.

628085 Integrated Flexibility and Controllability Analysis for the Continuous Pharmaceutical Manufacturing Process

Abstract:

Batch production mode has been dominant in the pharmaceutical industry during the past several decades. However, it has multiple disadvantages such as large storage volume requirements, strong scale-dependence and potential safety issues. Indeed, the shift from the conventional batch manufacturing to system to the continuous manufacturing method is gaining momentum. The advantages of Continuous Pharmaceutical Manufacturing (CPM) include an increase in efficiency of productivity and economy, maximized automation by the unit operations interconnection, enhanced product quality assurance and safety by a continual automated monitory of process[1].

One of central elements for the successful operation of CPM plants is whether the designed plant can be guaranteed to satisfy the specifications for the entire range of potential uncertain parameter values involved. From the perspective of process system engineering, it is essential to conduct the operability analysis for the CPM to guarantee the desirable dynamic performance under uncertainties. Indeed, flexibility and controllability are two crucial concerns of the operability analysis of chemical plants[2]. Flexibility guarantees feasible steady-state operation over a range of uncertain operating conditions and denotes static resiliency. Controllability reflects the attainable operation of a given process under dynamic operating conditions[3]. Traditionally, flexibility analysis focuses on processes that are operated at steady state within an uncertainty range, without taking into account the real-time control and dynamic performance of real chemical process. Flexible region may not ensure good performance under dynamic operating conditions.

This paper proposes a framework for the integration of process flexibility and controllability analysis for CPM process. We applied Relative Gain Array (RGA) as an indicator for controllability analysis which is incorporated into the flexibility analysis framework. A mixed integer nonlinear programming (MINLP) model is then formulated. Two cases including complicated micro-reactors to continuously produce Ibuprofen and Artemisinin are carried out to validate the proposed framework.

References

1. Domokos, A.; Nagy, B.; Szilágyi, B.; Marosi, G.; Nagy, Z. K., Integrated Continuous Pharmaceutical Technologies—A Review. *Organic Process Research & Development* **2021**.
2. Bahri, P. A.; Bandoni, J. A.; Romagnoli, J. A., Integrated flexibility and controllability analysis in design of chemical processes. *Aiche Journal* **1997,** *43* (4), 997–1015.
3. Yuan, Z.; Chen, B.; Zhao, J., An overview on controllability analysis of chemical processes. *AIChE Journal* **2011,***57* (5), 1185-1201.

625584 Model-Based Investigation of Upstream CHO Cell Culture Process for Production of Monoclonal Antibodies with Desired N-Linked Glycosylation

Abstract:

N-linked glycosylation is post-translational protein modification that takes place during synthesis of monoclonal antibodies (mAbs) [1]. N-linked glycan composition or glycan index, including total afucosylation, high mannosylation, fucosylation, and/or galactosylation indexes, are often critical quality attributes (CQAs) of mAb based therapeutic proteins [2]. Researchers have found controlling the cell culture media components such as uridine, galactose, manganese, and ammonia concentration can tweak the glycosylation composition with the desired CQA levels [3-5]. During bioreactor operations, many process parameters or operating conditions can also affect protein glycosylation such as temperature, pH and dissolved oxygen [6-8]. However, the effect of process parameters on mAb glycosylation can vary significantly between different cell lines. To better understand the protein glycosylation process and further optimize the bioreactor operations, process modeling can be used to support quality by design (QbD) by correlating critical process parameters with critical quality attributes and further defining a design space for process operations.

We have investigated the effect of temperature and pH on protein glycosylation for a model Herceptin biosimilar (or Trastuzumab) production using a CHO cell culture cultivation in an upstream fed-batch bioreactor. Mechanistic kinetic model was developed based on the experimental data obtained under different temperature and pH combinations. The mechanistic model contains two parts, including an unstructured cell culture model and a structured intracellular model. For our unstructured cell culture model, viable cell density, glucose, ammonia, and protein concentrations are captured by empirical equations. In the structured model, the intracellular kinetics of biochemical reactions cascade of the N-linked glycosylation process are simulated. The kinetic model is used to understand the effect of temperature and pH on the bioprocess. Finally, feasibility analysis with surrogate-based adaptive sampling is applied to determine a bioreactor operation design space. In this combined experimental and computational study, we have determined how adjusting the bioreactor operating conditions can achieve critical quality attributes of the targeted product. Ongoing work will implement integration with a control system to achieve real-time process control for mAb glycosylation.

References

1. Akmal, M. A.; Rasool, N.; Khan, Y. D. J. P. o., Prediction of N-linked glycosylation sites using position relative features and statistical moments. **2017,** *12* (8), e0181966.
2. Szekrenyes, A.; Szigeti, M.; Dvorakova, V.; Jarvas, G.; Guttman, A., Quantitative comparison of the N-glycosylation of therapeutic glycoproteins using the Glycosimilarity Index. A tutorial. *TrAC Trends in Analytical Chemistry* **2020,** *122*, 115728.
3. Gramer, M. J.; Eckblad, J. J.; Donahue, R.; Brown, J.; Shultz, C.; Vickerman, K.; Priem, P.; van den Bremer, E. T.; Gerritsen, J.; van Berkel, P. H., Modulation of antibody galactosylation through feeding of uridine, manganese chloride, and galactose. *Biotechnol Bioeng* **2011,** *108* (7), 1591-602.
4. Kotidis, P.; Jedrzejewski, P.; Sou, S. N.; Sellick, C.; Polizzi, K.; del Val, I. J.; Kontoravdi, C., Model-based optimization of antibody galactosylation in CHO cell culture. **2019,** *116* (7), 1612-1626.

5. Ehret, J.; Zimmermann, M.; Eichhorn, T.; Zimmer, A., Impact of cell culture media additives on IgG glycosylation produced in Chinese hamster ovary cells. *Biotechnol Bioeng* **2019,** *116* (4), 816-830.
6. Villiger, T. K.; Scibona, E.; Stettler, M.; Broly, H.; Morbidelli, M.; Soos, M., Controlling the time evolution of mAb N-linked glycosylation - Part II: Model-based predictions. **2016,** *32* (5), 1135-1148.
7. Sou, S. N.; Sellick, C.; Lee, K.; Mason, A.; Kyriakopoulos, S.; Polizzi, K. M.; Kontoravdi, C., How does mild hypothermia affect monoclonal antibody glycosylation? *Biotechnol Bioeng* **2015,** *112* (6), 1165-76.
8. del Val, I. J.; Kontoravdi, C.; Nagy, J. M., Towards the implementation of quality by design to the production of therapeutic monoclonal antibodies with desired glycosylation patterns. *Biotechnol Prog* **2010,** *26* (6), 1505-27.

624781 Next-Generation Vaccines and Therapeutics: Towards Resilient Pharmaceutical Supply Chains

Abstract:

The promising clinical outcomes of genetic engineering have paved the way to curing life-threatening diseases, including cancers. Several innovative gene therapy drugs have already been launched over the last few years and this rapid growth is forecasted to continue with the U.S. Food and Drug Administration (FDA) anticipating the approval of 10-20 cell and gene therapy products per year by 2025 [1]. Approximately 70% of the cell and gene therapies in the clinical trials utilise viral vectors for gene transfer [2]. The versatility of viral vectors has also emerged in preventive healthcare, with their recent approval as vaccine platforms (e.g. Vaxzevria (AstraZeneca) and the Janssen COVID-19 vaccine) [3][4].

Given the high demand and limited manufacturing capacity, viral vector manufacturers are asked to make decisions on how to best utilise and expand the available manufacturing resources in the presence of numerous demand uncertainties. Some of the challenges are related to production and capacity planning under uncertain clinical trial outcomes and dosage requirements. Furthermore, manufacturers of viral vector-based COVID-19 vaccines tackle the hurdles of rapid and in-risk manufacturing, transitioning to large-scale deployment of novel vaccine products in record times. In this case, SARS-Cov-2 has clearly demonstrated the need for sophisticated tools that can assist responsive decision-making and support planning under uncertainty.

This work presents a novel mixed-integer linear programming (MILP) model for the design and optimisation of viral vector supply chains (Figure 1). An array of applications for viral vectors and their respective demand scales are considered, thereby capturing the emerging paradigm shift from *one-size-fits-all* manufacturing and distribution to targeted healthcare [5]. The framework relies on techno-economic analyses of well-established viral vector manufacturing protocols, which are modelled and simulated in SuperPro Designer. The formulated optimisation model identifies good candidate supply chain structures, capacities and production plans for multi-product multi-suite facilities, distribution plans and transportation flows. For this, we consider (i) a set of products and demand scenarios, (ii) capital and operating manufacturing expenditures for a range of production scales, as well as storage and transport costs for sensitive formulated products, (iii) candidate locations for network nodes, namely primary, secondary manufacturing and distribution centres. The resilience of the strategic decisions is then assessed under demand uncertainty through comparisons of network structures with respect to (i) cost, (ii), scalability and (iii) flow time.

References

[1] **U.S. Food and Drug Administration,** "Statement from FDA Commissioner Scott Gottlieb, M.D. and Marks, M.D., Ph.D., Director of the Centre for Biologics Evaluation and Research on new policies to advance development of safe and effective cell and gene therapies", https://www.fda.gov/news-events/press-announcements/statement-fda-commissioner-scott-gottlieb-md-and-peter-marks-md-phd-director-center-biologics, (2019).

[2] **Ginn, S., Amaya, A. K., Alexander, I. E., Edelstein, M., Abedi, M.R.,** "Gene therapy clinical trials worldwide to 2017: An update", *The Journal of Gene Medicine,* **20**(5), e3015, (2018).

[3] **European Medicines Agency,** "Vaxzevria (previously COVID-19 Vaccine AstraZeneca)", https://www.ema.europa.eu/en/medicines/human/EPAR/vaxzevria-previously-covid-19-vaccine-astrazeneca#overview-section, (2021).

[4] **U.S. Food and Drug Administration**, "Janssen COVID-19 Vaccine", https://www.fda.gov/emergency-preparedness-and-response/coronavirus-disease-2019-covid-19/janssen-

covid-19-vaccine, (2021).

[5] **Sarkis, M., Bernardi, A., Shah, N., Papathanasiou, M.,** "Emerging Challenges and Opportunities in Pharmaceutical Manufacturing and Distribution", *Processes*, **9**(3), 457, (2021).

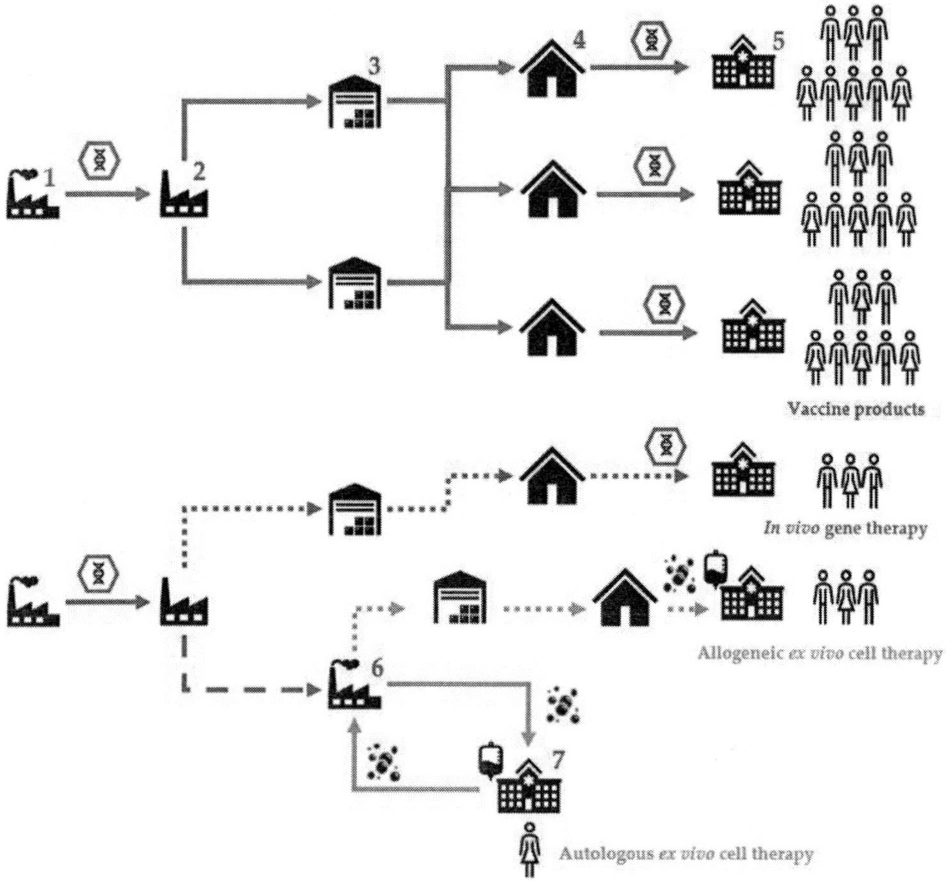

Figure 1: General schematic of supply chains for viral vector applications, comprising (1) primary manufacturing, (2) secondary manufacturing/fill-to-finish, (3) regional storage, (4) district storage, (5) hospitals/administration points as well as (7) specialised leukapheresis sites, (6) cell therapy manufacturing facilities.

624019 Systematic Decomposition & Evaluation of a Process Design Space for Monoclonal Antibody (mAb) Manufacturing

Abstract:

Three of the major trials within drug development and commercialization are phase 1 to 3 clinical trials. As these trials proceed product-process understanding increases and these are utilized via a feedback loop to the different development areas. The complex feedback loop consists of knowledge from input and output material properties (product related) and unit operations performance (process related). Monoclonal antibodies (mAbs) are of importance because they are proteins that are purposely engineered to treat various severe diseases for example cancer or hemophilia (Wolf Pérez et al., 2019). A typical pharmaceutical process for a specific active pharmaceutical ingredient (API) is batch-based because batch processes provide the flexibility of quality control at different intermediate steps and batch tracing inside the process boundaries and outside, that is, the product supply chain. Of interest, whether a continuous or batch process, is the evaluation of technologies that have a positive impact on process performance. Therefore, it is of interest to systematically evaluate relevant technologies for improving process performance and sustainability beyond high level "what-if analyses" to provide the foundation for advanced process understanding, design and optimization (Yang et al., 2020).

The mAb platform process to be investigated is pre-selected to be a batch process and therefore, any pre-selected flowsheet model for representation of the process is dynamic. Therefore, in analyzing upscale data for the platform process (lab and/or pilot), a systematic approach where multiple ideas associated with evaluating new technologies with respect to mass throughput and time execution is of importance. For the platform process, based on the large number of alternative placements of new technologies (integer decisions) and estimation of the input/outputs (continuous decisions) a systematic approach consisting of two layers of decisions is proposed. The decomposition layer addresses the integer decision problem, that is, placement of the new technology and the evaluation layer; having fixed the placement; addresses the continuous decision, that is, the expected gain in process performance, capacity and sustainability. This is performed using a simulation-based approach of material flow analysis with embedded non-linear process behavior.

The objective of this presentation is to present the systematic method for mAb process design space generation and screening. First, the problem and method will be presented. Second, the concept of discrete-event simulation will be introduced and how this can be used within the method for decomposition and evaluation of the design space. Finally, the method will be exemplified through an example of a promising technology for mAb manufacturing.

References

Wolf Pérez A. M., Sormanni P., Andersen J. S., Sakhnini L. I., Rodriguez-Leon I., Bjelke J. R., Gajhede A. J., De Maria L., Otzen D. E., Vendruscolo M. and Lorenzen N. 2014. In vitro and in silico assessment of the developability of a designed monoclonal antibody library. mAbs, 11(2), 388–400.

Yang, O., Qadan, M. & Ierapetritou, M. Economic Analysis of Batch and Continuous Biopharmaceutical Antibody Production: a Review. 2019. Journal of pharmaceutical innovation, 15, 182–200.

621647 Keynote Talk: Integrated Quality By Design in (Bio)Pharmaceutical Manufacturing

Abstract:

This presentation describes a Quality-by-Design approach for an accelerated process development workflow for the design of integrated (bio)pharmaceutical manufacturing systems from drug substance to drug product. This approach draws R&D and management into closer alignment and execution while defining product critical quality attributes, the operational design space, the manufacturing process, and the control strategy by integrated use of the ICH quality quartet – namely Q8(R2), Q9, Q10, and Q11. The workflow employs process intensification, micro-scale technologies, modular unit operations with integrated process control and monitoring systems, systems integration, and full plant automation. The presentation describes how to best develop and transfer knowledge between steps in the workflow via mechanistic models, data analytics, and machine learning. The approach is illustrated in more than a half dozen small- and large-molecule applications, including where process equipment was first designed digitally using first-principles models, then constructed and implemented experimentally, with the experimental results confirming model predictions. These applications include a wide variety of upstream, downstream, and formulation unit operations for small-molecule and biological drug manufacturing, as well as the design and operation of fully automated end-to-end manufacturing systems.

624003 Keynote Talk: An *in silico* approach for Monoclonal Antibody (mAb) Process Research & Early Development

Abstract:

Monoclonal antibodies (mAbs) are a type of biopharmaceutical that can be used to treat various diseases, for example, hemophilia and other serious diseases (Wolf Pérez et al., 2019). Improvements for the development of mAbs can be categorized as follows, product related improvements, for example, expression, solubility etc. and process related improvements, for example, novel flowsheets, new technologies (unit operations) etc. The synergistic effect of product-process decisions is also of importance because it addresses the integrated problem, that is, how can process design decisions be accounted for in the product design phase. Consider the example, given a specified mAb, purify the mAb using a chromatography column and selected eluent at a predetermined concentration. If the resulting aggregated pool volume for a finite time horizon for a pilot and/or industrial process requires a non-standard tank size that increases capital investment, how can this be a priori mitigated? If product and process design is performed sequentially then the design decision is unchanged however, if product-process design is performed integrated then, additional technologies can be evaluated for up-concentration. Therefore, the process is sized to the required scale and not oversized which has both capital investment and time-to-market implications.

Production planning-scheduling (PPS) is a field within operations research that solves the dynamic problem as to what to produce when (Harjunkoski et al., 2014). It can be performed using optimization-based methods and simulation-based methods, for example, discrete event simulation. The former differs from the latter because it provides an optimal solution for a set of decisions however, the latter provides the opportunity to deep dive into the details of the plan-schedule. Therefore, the symbiosis between the two approaches can be explained as follows. The former generates the optimal plan-schedule and the latter simulates how this is executed. In other words, the latter acts as a digital version of the process, a so-called digital twin. Therefore, we can use discrete event simulation to also evaluate the impact of different technologies and various design decisions on pharmaceutical processes, evaluate economic and sustainability impacts and improve decision making a priori to final investment (Yang et la., 2020).

The objective of this presentation is to present an *in silico*, systematic framework for multiscale process design and idea evaluation for biopharmaceutical products. A case study applied to mAbs will be presented. The method will be presented, and its application exemplified through idea evaluation related to new technologies for mAb production. Finally, the screening of the evaluation space using a combination of economic and sustainability metrics will be presented.

References

Harjunkoski I, Maravelias CT, Bongers P, Castro PM, Engell S, Grossmann IE, Hooker J, Méndez C, Sand G and Wassick J. 2014. Scope for industrial applications of production scheduling models and solution methods. Computers and Chemical Engineering, 62, 161–193.

Wolf Pérez A. M., Sormanni P., Andersen J. S., Sakhnini L. I., Rodriguez-Leon I., Bjelke J. R., Gajhede A. J., De Maria L., Otzen D. E., Vendruscolo M. and Lorenzen N. 2014. In vitro and in silico assessment of the developability of a designed monoclonal antibody library. mAbs, 11(2), 388–400.

Yang, O., Qadan, M. & Ierapetritou, M. 2019. Economic Analysis of Batch and Continuous Biopharmaceutical Antibody Production: a Review. Journal of pharmaceutical innovation, 15, 182–200.

631710 A Deep Learning Approach for Chord Length Distribution Modeling of Two-Dimensional Crystals

Abstract:

Despite significant advances in process analytical technology, online monitoring of particle size distributions (PSDs) in industrial crystallizations remains a challenge. Laser backscattering using focused beam reflectance measurement (*FBRM*) has been widely shown to be a promising technique for online, but indirect measurement of PSDs [1]. In fact, FBRM measures chord length distribution (CLD), which does not provide a physical meaning of crystal size (e.g., [2,3]).

Towards converting CLDs to PSDs to enhance process monitoring capabilities, in this talk, we will present a deep learning approach for automatically converting simulated PSDs to CLDs which resemble the ones measured by FBRM. Deep learning and deep learning inspired methods have been recently applied to map the relationship between CLD and PSD [2,3,4], though without fully exploring their potential and limitations. In particular, we use convolutional neural networks (CNNs) [5], which are frequently used for image classification as they excel at identifying spatial patterns and relationships within data. By discretizing the PSD into finite bins, the multi-variate PSD can be effectively treated as an "image". As a result, each PSD is converted into an image and serves as the input to the CNN, while conveying the information about the frequency, position, and spacing of each bin within the discretized distribution. We use Bayesian optimization (BO) [6] to automatically tune hyperparameters of the CNN model. Specifically, BO is used to determine the optimal number of filters to use in each of the convolutional layers, as well as the optimal CNN training parameters.

To demonstrate the proposed deep learning approach, we first use simulation data of CLDs and PSDs for needle-like crystals to train and validate a CNN model. We show that the CNN model performs with an industrially pragmatic accuracy. We then investigate the generalizability of the CNN using transfer learning techniques [7], with the goal of obtaining better predictions for "unseen" types of crystals not included in the training data, or in cases where there is a limited amount of data. In particular, we examine the transfer learning by using a CNN that is trained for needle-like crystals, the dense layer of which is then retrained using a limited dataset of platelet-like crystals, while the convolutional layers part that performs feature extraction remains "frozen". We demonstrate the effectiveness of the CNN model augmented with transfer learning using experimental data. The proposed deep learning approach for converting PSDs to CLDs, which relies only minimally on system specific measurements, often time-consuming and expensive, and can be particularly useful for enhancing process monitoring capabilities in industrial crystallization processes, as well as laboratory studies for elucidating crystallization kinetics.

[1] Simon, L. L., Pataki, H., Marosi, G., Meemken, F., Hungerbühler, K., Baiker, A. & Chiu, M. S. (2015). Assessment of recent process analytical technology (PAT) trends: a multiauthor review. *Organic Process Research & Development*, *19*(1), 3-62.

[2] Szilagyi, B., & Nagy, Z. K. (2018). Aspect ratio distribution and chord length distribution driven modeling of crystallization of two-dimensional crystals for real-time model-based applications. *Crystal Growth & Design*, 18(9), 5311-5321.

[3] Crestani, C. E., Bernardo, A., Costa, C. B. B., & Giulietti, M. (2021). An artificial neural network model applied to convert sucrose chord length distributions into particle size distributions. *Powder*

Technology, *384*, 186-194.

[4] Irizarry R., Chen A., Crawford R., Codan L., & Jöchen S. (2017). Data-driven model and model paradigm to predict 1D and 2D particle size distribution from measured chord-length distribution. *Chemical Engineering Science*, 164, 202-218.

[5] Bengio, Y., Goodfellow, I., & Courville, A. (2017). *Deep learning* (Vol. 1), Massachusetts, USA: MIT press.

[6] Snoek, J., Larochelle, H., & Adams, R. P. (2012). Practical Bayesian optimization of machine learning algorithms, *arXiv preprint arXiv:1206.2944*

[7] Tan, C., Sun, F., Kong, T., Zhang, W., Yang, C., & Liu, C. (2018). A survey on deep transfer learning, "*International conference on artificial neural networks*" (pp. 270-279). Springer, Cham.

622939 Handling and Solid-State Method Development for Improved Isolation of Molecules with Complex API Structures

Abstract:

The pharmaceutical industry in general will continually encounter larger and more complex molecules due to the need for more novel disease targets. With these highly flexible and larger molecular structures come a higher propensity for solid forms of the API with more atypical behavior, e.g. variable hydrates/solvates and pseudo-stable amorphous forms that display a continuum of energetic states, rather than discrete solid form transitions. New approaches and methodologies must be explored and used to better characterize and understand these molecules to ensure robust and consistent isolation of the desired form, and to inform regulatory authorities upon submission. Pharmaceutical case studies are presented describing how adapted methodologies are used to provide better understanding of complex solid forms with regard to solid-state properties and the isolation process. The utility of different advanced solid-state characterization tools such as VH/T-XRPD, TGA-MS, perfusion calorimetry, and Raman spectrometry are reviewed and discussed. In the first case study, the aforementioned characterization tools were used to guide development of a crystallization and isolation process for a variable hydrate/solvate that exhibited small subtle changes in crystallinity at different environments, which impacted the crystallization design and performance. The second case study describes an API that undergoes form change and partial amorphization during drying. The developed solid-state methods were able to quantify the extent of amorphous content and how it is directly related to desolvation behavior. The knowledge generated by these case studies not only supported advancement through development, but will also contribute to an effective presentation of the properties of these more complex forms to regulatory authorities.

AIChE 2021 abstract for verbal presentation

Title: **Hollow crystal formation through a novel ripening mechanism in crystallization**

Authors: Fredrik L. Nordstrom*, Yongjian Wang, Huayu Li, Michelle Raikes, Brian Linehan, John Robson

Affiliation: Material and Analytical Sciences, Boehringer-Ingelheim, Ridgefield, CT

*Email: fredrik.nordstrom@boehringer-ingelheim.com

Abstract:

A novel ripening mechanism in crystallization is presented that allows for crystals to undergo morphological changes over time at seemingly constant conditions. This ripening phenomenon is different from Ostwald ripening in terms of its thermodynamic driving force and applicability across different particle sizes. Salicylic acid was crystallized in the presence of simulated structurally similar impurities, which resulted in formation of solid solutions. The impurity entrapment was elevated initially as a result of secondary nucleation and decreased during the course of the crystallization. The final crystals were thus comprised of impurity gradients having dirty cores and cleaner outer edges. Confocal Raman imaging was used to confirm the variable impurity entrapment in individual crystals showing enrichment in the center and interior of the crystals. As higher impurity levels corresponded to elevated solubility, dissolution took place from the impure center of the crystals when the solution concentration approached the solubility of pure salicylic acid. This in turn led to formation of macrotubular crystals with clearly distinguishable holes and cavities. The ripening mechanism is demonstrated experimentally at different stages in a crystallization relevant to a typical pharmaceutical isolation process. The results show not only that hollow crystals can be designed and challenging impurities can be better rejected, but more importantly it demonstrates the fundamental and dynamic interplay between impurity entrapment, solubility and desaturation kinetics that takes place throughout a real crystallization process.

Fig 1. Hollow crystal of salicylic acid after ripening

623775 Crystallization Engineering for the Robust Manufacturing of Islatravir Using SINTAX (shear-induced nucleation and thermal annealing crystallization)

Abstract:

Islatravir (MK-8591) is a potent nucleoside reverse transcriptase translocation inhibitor in development for HIV treatment and prevention. We have developed a robust crystallization method that ensures the active pharmaceutical ingredient (API) meets the required purity and physical property attributes to enable sub-milligram doses and multiple formulation types. Our crystallization takes a crude API, containing high levels of biologic and chemical impurities, and produces an API with high purity for drug substance quality and consistent particle size for content uniformity and improved manufacturing.

Herein, we highlight the development and scale-up of a shear-induced nucleation and thermal annealing crystallization (SINTAX) that delivers islatravir with a controlled particle size distribution and impurity profile[1]. We emphasize lessons learned by applying SINTAX at the lab, pilot, and commercial scales. Across manufacturing scales, we demonstrate the robustness of SINTAX and the importance of solubility models and process analytical technologies (PAT). Altogether, we show the importance of crystallization engineering to control impurity rejection, particle characteristics, and processability for the development of a life-changing drug.

References:

1. Sirota, E., Kwok, T., et al. "Crystallization Process Development for the Final Step of the Biocatalytic Synthesis of Islatravir: Comprehensive Crystal Engineering for a Low-Dose Drug." *Organic Process Research & Development* 25.2 (2021): 308-317.

629233 Evaluation of Liquid-Liquid Phase Separation and Spinodal Boundaries Using a Continuous-Flow Microfluidic Mixer

Abstract:

Liquid-liquid phase separation (LLPS), also known as oiling out, refers to the process where the liquid droplets with high concentration of solute phase separate in a supersaturated solution. Oiling-out is a common phenomenon among hydrophobic pharmaceutical compounds. The spontaneous nature of the transformation of oil droplets into solids makes LLPS very hard to measure. As a result, it is considered an undesired phenomenon and is typically avoided during the crystallization. Therefore, the most critical step in process development is to obtain the phase diagram to ensure the crystallization does not occur within the LLPS zone. Here, we have used the previously designed continuous-flow microfluidic mixer to evaluate the LLPS phenomena of Beta-Alanine. In this study, we present the isothermal ternary phase diagram of beta-alanine in water and isopropyl alcohol (IPA) or ethanol.

All the following experiments are performed using an optical microscope to track and record the appearance and disappearance of oil droplets. The microfluidic mixer consists of four tangential inlets. Two of them are streams of a concentrated beta-alanine solution in water, one is pure solvent (water), and the last one is pure anti-solvent (IPA or Ethanol). The procedure will start by fixing the ratio of the anti-solvent to solvent stream. Then we will start with the highest flowrate for the pure solvent stream at a fixed ratio of anti-solvent. After that, the flow rate of beta-alanine streams is gradually increased while that for the pure solvent is gradually decreased. The composition at which the droplets appear in the mixture is considered the spinodal point. Once the oil droplets appear, the flowrate of solvent stream is gradually increased until the droplets disappear. This composition indicates the binode. Finally, these spinodal and binodal points are used to plot the ternary phase diagram to highlight the LLPS zone.

630606 Molecular Simulations Reveal Underlying Mechanism of Cooling and Antisolvent Crystallization to Predict the Polymorphism and Growth of Organic Crystals

Abstract:

Understanding the self-assembly of molecules driven by increased chemical potential is a fundamental aspect of materials science. Crystallization of organic molecules is one such process where the increase in supersaturation is used to obtain the crystal of the desired molecule. However, the highly stochastic nature of molecular motion and the presence of multiple energy minima in the crystal energy landscape makes it difficult to predict the outcome of crystallization. This problem is more significant for the pharmaceutical industries, where multiple energy minima in the crystal energy landscape lead to polymorphism. To understand the molecular events which lead to polymorphism, we have employed molecular simulations of cooling and antisolvent crystallization processes. The simulations reveal that the rapid exchange of solvent molecules in the solvation shell and the configuration of the solvent molecules around different functional groups of the organic molecule can be used to predict the polymorphic form obtained at the end of the crystallization. In cooling crystallization simulations of glutamic acid molecules, the density of the solvent molecule in the solvation shell influences the polymorph formation. However, in the case of antisolvent crystallization of Histidine in the mixture of Water and Ethanol molecules, the rupture of the solvation shell due to the interaction with the Ethanol molecule influences the polymorphism. Furthermore, knowledge of per-molecule energy contribution in the solvation shell allows calculating the growth rates of the crystals, which can be used to control the morphology of the crystals. Overall, the knowledge of solvation shell dynamics explains and unifies the traditional empirical observations and opens up possibilities for better control strategies for crystallization.

623812 Cephalexin and Amoxicillin Crystal Shape Modification By Manipulating the Supersaturation and Wet Milling

Abstract:

Cephalexin and amoxicillin belong to the family of beta-lactam antibiotics and are among the most widely used small-molecule drugs. Both compounds are sparingly soluble in water, the solvent in their enzymatic synthesis process, and form hydrated crystals under typical synthesis conditions [1]. Unfortunately, both cephalexin and amoxicillin crystallization processes are reported to produce needle-like crystals with high aspect ratios, which can lead to significant agglomeration, long filtration times, and poor downstream processability [2]. There have been several efforts to address the above issue and to modify the crystal habit and shape in these systems. For example, it has been suggested that seeding and controlling the supersaturation generation through controlling either the cooling rate [3], synthesis reaction rate [4], or feeding rate [5] can be used to improve the shape of cephalexin monohydrate crystals.

In the present work, we examine the impact of wet milling on the crystal shape, aspect ratio, and size distribution in cephalexin and amoxicillin crystallization processes. Wet milling is commonly used to reduce the crystal size in different processes and to tighten the size distribution. Our results suggest that wet milling not only reduces the mean size of cephalexin monohydrate crystals, but it also leads to further improvements in crystal aspect ratio during the enzymatic synthesis and crystallization process. Based on a systematic study of different milling parameters, we will discuss the impact of milling on crystals of both cephalexin and amoxicillin. Furthermore, we investigate potential improvements in the aspect ratio of the final crystals by combining the milling with a series of pH manipulation steps where small, needle-like crystals are dissolved by periodically increasing the system pH (leading to higher solubility), giving a way to produce larger crystals with improved aspect ratios (Figs. 1A and 1B). The impact of the improved aspect ratio on the filtration rate of both crystals will also be discussed (Fig. 1C).

Fig. 1. Example of improvement in the aspect ratio of cephalexin monohydrate crystals by applying a series of pH swings (A), or periodic milling (B). Both cases lead to the formation of more platelike crystals compared to initially needle-like crystals. Example of the impact of aspect ratio (or size) on filtration rate (C).

Cited literature

[1] McDonald, Matthew A., Grant D. Marshall, Andreas S. Bommarius, Martha A. Grover, and Ronald W. Rousseau. "Crystallization kinetics of cephalexin monohydrate in the presence of cephalexin precursors." *Crystal Growth & Design* 19, no. 9 (2019): 5065-5074.

[2] Perini, Giulio, Carlos Avendaño, William Hicks, Anna R. Parsons, and Thomas Vetter. "Predicting filtration of needle-like crystals: A Monte Carlo simulation study of polydisperse packings of spherocylinders." *Chemical Engineering Science* 230 (2021): 116151.

[3] Ostergaard, Iben, Heidi Lopez de Diego, and Haiyan Qu. "Crystallization of cephradine polymorphs and hydrates from mixed solvents of methanol and water." *Chemical Engineering & Technology* 42, no. 7 (2019): 1512-1518.

[4] McDonald, Matthew A., Grant D. Marshall, Salami, Hossein, Andreas S. Bommarius, Martha A. Grover, and Ronald W. Rousseau. "Crystallization Optimization By Enzymatic Control of a Reactive Crystallization". AIChE annual meeting, San Francisco (2020)

[5] Li, Mingchen, Zeren Shang, and Baohong Hou. "Optimizing the Aspect Ratio of Cephalexin in Reactive Crystallization by Controlling Supersaturation and Seeding Policy." *Transactions of Tianjin University* 25, no. 4 (2019): 348-356.

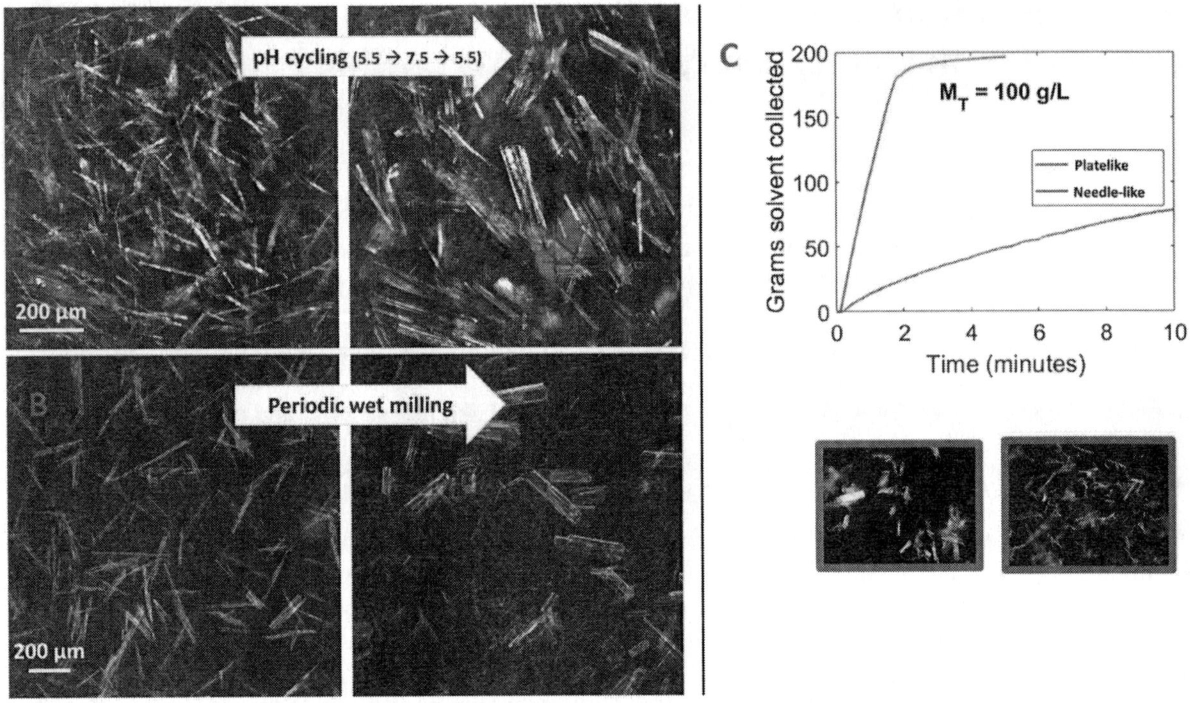

A
pH cycling (5.5 → 7.5 → 5.5)
200 µm
B
Periodic wet milling
200 µm
C
M_T = 100 g/L
Platelike
Needle-like
Grams solvent collected
Time (minutes)

625129 Accelerated Early-Stage Enabling API Crystallization Process Development and Scale-up

Abstract:

Development of crystallization, filtration/washing, and drying processes is an essential cornerstone of delivering quality Active Pharmaceutical Ingredient (API) in the pharmaceutical industry. During early-stage development, rapid timelines and the need to generate high quality API to enable GLP toxicity studies and Phase I clinical trials necessitate rapid and targeted data rich experimentation to develop robust and scalable processes. While early-stage crystallization development often falls upon the chemists, early intervention with an engineering and scalability-focused mindset can significantly reduce the development burden later in the pipeline. Such development efforts on Compound A, and the enabling data-rich workflow leveraged to yield a robust crystallization, filtration, and drying process is detailed in this work. A rapid 6-week development period, followed by a successful scale up from gram to kilo scale over a subsequent two-week period enabled a GLP toxicity study to proceed for the early-stage Compound A.

Swift crystallization solvent selection was aided by a combination of in-silico modeling using COSMO quantum chemistry software in conjunction with experimental UHPLC based solubility measurements. Selection of a suitable solvent-antisolvent system was further informed using DynoChem®, which was leveraged to generate full solubility curves over a range of temperatures and compositions with limited experimental data. Leveraging these tools is often essential in early-stage process development due to limited material availability. A favorable solubility profile for a given composition was identified, and a simple cooling crystallization was designed to afford high yield and purity. Material lean experiments using a variety of process analytical technologies (PAT) were designed to determine the metastable zone width and explore seeding conditions (seed load and supersaturation ratio) to ensure consistent crystal form control while avoiding oiling out. A complex form landscape during filtration and drying was mapped with DSC/TGA/MS and PXRD and navigated through to ensure the physical properties of the API were preserved during filtration, washing, and drying. Carefully tailored displacement washes were developed and leveraged to avoid crystal form change. A terminal wet milling and thermocycling process was also developed to control crystal surface area in conjunction with appropriate drying process understanding to achieve target specifications in the lowest cycle time. The process was successfully scaled-up to kilo scale to enable the GLP toxicity study.

Disclosures:

All authors are employees of AbbVie and may own AbbVie stock. AbbVie sponsored and funded the study; contributed to the design; participated in the collection, analysis, and interpretation of data, and in writing, reviewing, and approval of the final publication.

630526 Development and Scale-up of a Crystallization Process for a Kinetically-Unfavorable Polymorph

Abstract:

ABSTRACT: This work describes the development and scale-up of a crystallization process to isolate a thermodynamically stable but kinetically unfavorable polymorph of a small organic agrochemical prone to oiling. The process involves simultaneous addition of solution and antisolvent into a slurry of the desired polymorph, similar to semi-batch, reverse addition antisolvent crystallization.[1] For this system, the simultaneous addition process is more robust to feed variation than a traditional antisolvent process and affords suitable polymorph control, recovery, cycle time, and purity. However, the process produces a thick, but shear-thinning slurry of fine, needle-like particles that is challenging to mix well. Mixing analysis tools, including both the Fluid Mixing Processes (FMP) Excel Design Guide and computational fluid dynamics simulations were utilized to enable successful scale-up of this mixing-sensitive process from lab to pilot plant.

[1] R. Pena, C.L. Burcham, D.J. Jarmer, D. Ramkrishna, Z.K. Nagy, "Modeling and optimization of spherical agglomeration in suspension through a coupled population balance model," *Chemical Engineering Science,* 167 pp. 66-77, 2017.

632628 Acid-Catalyzed Esterification Governs the Chain Elongation and the Oriented Attachment in Cof-5 Synthesis

Abstract:

Covalent-Organic Frameworks (COFs) have found application in a wide range of fields such as catalysis, membrane separations, adsorption, and drug delivery due to their high symmetry, porosity, crystallinity, and large surface area. However, long timescales of the synthesis, difficulties in characterization, formation of amorphous intermediate, and the limited understanding of the underlying reaction network have prevented the controlled synthesis of COF crystals. Furthermore, the existing solvothermal synthesis techniques are difficult to reproduce. To overcome these limitations, improve reproducibility, and obtain quantitative rate contributions of different polymerization phases, we employ experimental studies of a wide range of operating conditions to synthesize COF-5 crystals. The yield and grain sizes of the COF crystals were characterized using reliable and reproducible ex-situ techniques such as Fourier Transformed Infrared Spectroscopy (FTIR) and powdered X-ray diffraction (P-XRD). Optimization of the temperature, reactant concentrations, type of solvents, and catalysts have provided significant mechanistic and resulted in low reaction times, high yields, and larger particle sizes. The insights gained from the comprehensive experimental studies allowed the development of a microkinetic model which simulates millions of nodes in the COF reaction network. The error-minimized parameters in the microkinetic models reproduce the experimental results of COF-5 yield and grain sizes in a predictive manner. Furthermore, the microkinetic approach also validates the proposed reaction mechanism obtained from the experimental studies. Results suggest that the acid-catalyzed esterification governs the chain elongation and the oriented attachment of COF-5 crystals.

621619 Enabling a Selective Dissolution Scheme for the Removal of Fines in Crystallization with a Hydrocyclone: Modeling and Experimental Validation

Abstract:

Effective methods to tune the particle size distribution (PSD) are of major interest in the field of crystallization, with the aim of facilitating the downstream processing, improving the mechanical properties of the powder, and, in the case of crystalline pharmaceuticals, modulating the absorption rate of the active ingredient in the human body.

Fines, i.e., crystals of a few tenths of micrometers in size, are often an undesired portion of the PSD, as they are responsible for extended filtration times. For their removal, temperature cycling is usually applied, consisting of one or more repetitions of a partial dissolution step followed by a growth step [1,2]. Temperature cycling has the advantage of being a very simple procedure; however, depending on the PSD and on the compound-specific behavior of crystallization, its efficiency may be limited, and the outcome may not be well predictable. Two factors limit the efficiency and predictability of temperature cycling. First, the partial dissolution step is non-selective, as it acts on the whole suspension and also affects larger crystals, thus decreasing the productivity; second, the evolution of the PSD strictly depends on the kinetics of dissolution and growth, which are generally unknown, and require tedious lab-work to be evaluated.

To address both shortcomings of temperature cycling, we have proposed and analyzed in detail through modeling an alternate process featuring a hydrocyclone [3], referred to as a selective dissolution process and outlined in Figure 1a. Hydrocyclones are common and inexpensive devices employed in several fields of chemical engineering, where they can act as classifiers to separate particles based on their size [4]. In the selective dissolution process, first the hydrocyclone separates finer from coarser crystals, then the suspension containing the fines is heated, so that only the particles in the targeted size range are dissolved. Further, the separation in the hydrocyclone depends on its design and operating conditions, and on the physical properties of the suspension, thus making the process less reliant on the kinetics of crystallization.

In this contribution we present both the main findings of the above-cited modeling study and the results of novel validation experiments, highlighting the trends and trade-offs that characterize the operation of a hydrocyclone and their effect on the overall performance of the crystallization process. Particular attention is devoted to process design, and a pathway for the selection of a suitable hydrocyclone design for a given target of fines removal is outlined.

The experimental campaign has been performed with β L-glutamic acid in water. 3D printing has been exploited to manufacture the hydrocyclones in-house (shown in Figure 1b), so as to enable superior flexibility in the design of units with different geometries. The experimental results demonstrate the advantage of the selective dissolution process over standard temperature cycling, and validate the modeled trends. Finally, the experimental data has been used to fine-tune the model predicting the separation performance of the hydrocyclone, which can be therefore used reliably for process design.

[1] Nagy, Z. K.; Aamir, E.; Rielly, C. D. Internal Fines Removal Using Population Balance Model Based Control of Crystal Size Distribution under Dissolution, Growth and Nucleation Mechanisms. *Growth Des.* **2011**, *11*, 2205–2219.

[2] Salvatori, F.; Mazzotti, M. Manipulation of Particle Morphology by Crystallization, Milling, and Heating Cycles: Experimental Characterization. *Eng. Chem. Res.* **2018**, *57* (45), 15522–15533.

[3] Binel, P.; Mazzotti, M. A Selective Dissolution Process Featuring a Classification Device for the Removal of Fines in Crystallization. *Eng. Chem. Res.* **2021**, *60* (1), 614–628.

[4] Svarovsky, L. Hydrocyclones. In *Solid-Liquid Separation*; Svarovsky, L., Ed.; Butterworth-Heinemann: Oxford, U.K., 2001; pp 191–245.

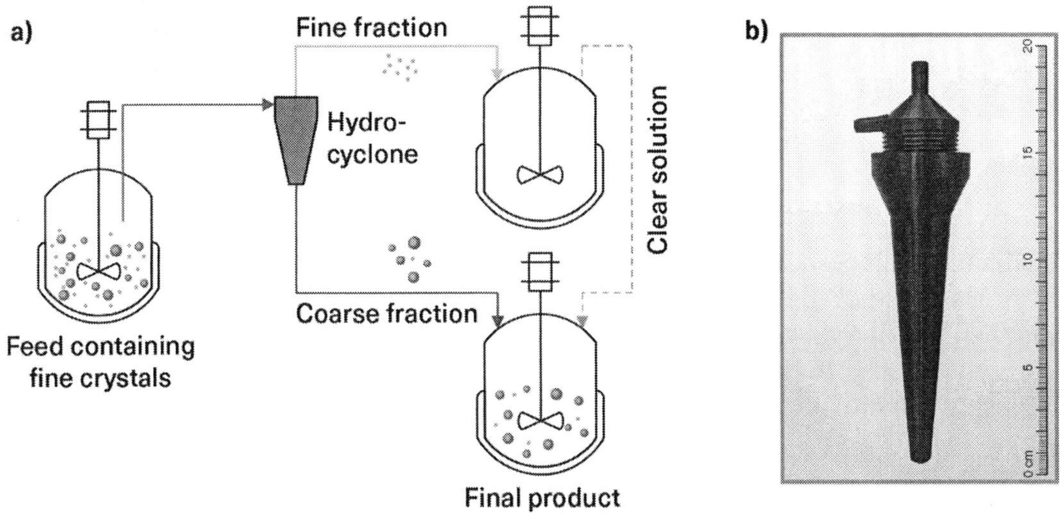

Figure 1. (a) Scheme of the selective dissolution process. The feed suspension is pumped through the hydrocyclone, which splits it into two streams; one stream contains most fine crystals while the other contains most coarse crystals. The stream containing the fines is collected in a vessel where complete dissolution occurs. The clear solution is then combined with the suspension containing the coarse crystals, and the supersaturation is depleted until equilibrium, i.e., the end of the process. (b) Photograph of a 3D printed hydrocyclone. For improved printing performance and flexibility, it is composed of a top cylindrical part and a bottom conical part. The top part is screwed into the bottom part, ensuring a tight fit and a smooth transition between the two.

627313 Crystallization Modeling of a Pharmaceutical Compound for Digital Twin Based in-Silico Optimization with Experimental Validation

Abstract:

The design of crystallization processes to achieve robust critical quality attributes (QCAs) is a practical challenge in the pharmaceutical industry. The current industrial standard still highly depends on Quality-by-Design (QbD), which requires intensive experimentation using traditional design of experiments (DoE) to find the operation conditions in the experimental space that satisfy the QCAs. With the current emerging Industry 4.0 practices the emerging paradigm has become the introduction of digital twins, which provide a predictive tool to be used in the optimization of the crystallization processes, or for in-silico DoE studies.[1]

In this work, we present the development and validation of a digital twin for the crystallization process of a pharmaceutical API (Compound A), from Takeda Pharmaceuticals International Co., which forms needle-like crystals. Several model training and validation experiments were performed with temperature cycling to promote the growth of the crystals for kinetic studies. For data acquisition and system monitoring during these experiments, process analytical technology tools (PAT) were used, including Mettler Toledo's ParticleTrack G400 for in-line chord length distribution (CLD), for detecting the crystallization events such as nucleation and dissolution, as well as for the crystal count measurement, the Zeiss MCS621 ATR-UV/Vis spectrophotometer for concentration measurement, and the Mettler Toledo's ParticleView v19 for in-situ microscope images. Additionally, Malvern's Mastersizer 3000 Hydro MV was used as an off-line characterization tool to measure crystal size distribution (CSD) and the data from the PAT tools were directly used in the parameter estimation step.

Two novel features were introduced during the parameter estimation for digital twin development. First, to make the model parameter estimation more robust, the focused beam reflectance measurement (FBRM) data was directly used in the parameter estimation without further transformations. Secondly, a novel size dependent growth (SDG) rate expression was used in the model to capture the CSD dynamics considerably better than the standard SDG rate models. The developed digital twin includes secondary nucleation, SDG, and dissolution mechanisms, which were identified by the designed training experiments and the objective that required thermocycling for the desired CQAs. The aim was to grow the crystals as large as possible without exceeding a batch time of 40 h while having product crystals with bulk density values larger than 0.25 g/mL. For this purpose and from previous experience, the training and validation experiments were designed to have temperature cycles.[2,3] The developed model was used for in silico DoE and process optimization and the simulation results were validated experimentally, demonstrating the benefits of model-based digital design for crystallization process development. In addition, the benefits of having a digital twin for in-silico optimization were demonstrated by comparing the findings from an experimental DoE work on Compound A to the ones from the optimized profiles, which were obtained by using the digital twin. The results demonstrated that the developed model was accurate and predictive enough to validate the general conclusions related to the required key features of the optimal temperature profiles identified by the experimental DoE. These results indicated that in order to obtain larger crystals with narrower CSD, multiple temperature cycles should be implemented at higher temperature ranges (50-65 °C). It was also shown that having a higher range for cycling with multiple cycles has a larger effect on product crystals size and bulk density than changing the seed type and loading.

The conclusions of the in-silico DoE and digital design were validated by experiments, demonstrating that the model was able to identify the key features of the optimal operating conditions for the crystallization of Compound A, when the aim was to produce large crystals by implementing temperature cycles. The orders of magnitude difference in the time required to reach this conclusion by in-silico and experimental DoE also demonstrated the benefits of having a digital twin of the process that can enable rapid crystallization process development by using a digital design framework.

Declaration of Competing Interest

The authors declare that they have no known competing financial interests or personal relationships that could have appeared to influence the work reported.

References:

1. Su, Q. *et al.* A perspective on Quality-by-Control (QbC) in pharmaceutical continuous manufacturing. *Comput. Chem. Eng.* **125**, 216–231 (2019).
2. Jiang, M. *et al.* Modification of crystal shape through deep temperature cycling. *Ind. Eng. Chem. Res.* **53**, 5325–5336 (2014).
3. Simone, E., Klapwijk, A. R., Wilson, C. C. & Nagy, Z. K. Investigation of the Evolution of Crystal Size and Shape during Temperature Cycling and in the Presence of a Polymeric Additive Using Combined Process Analytical Technologies. *Cryst. Growth Des.* **17**, 1695–1706 (2017).
4. Mydlarz, J. & Jones, A. G. On the estimation of size-dependent crystal growth rate functions in MSMPR crystallizers. *Chem. Eng. J.* **53**, 125–135 (1993).

624113 Thermoplastic Copolyesters: A Novel Solution for 3D-Printed, Implantable Drug Delivery Devices

Abstract:

In the last decade, individualized, patient-centered formulations have gained attention due to the emerging 3D-printing technologies [1]. One such technology is fused filament fabrication (FFF), which is currently mainly applied to oral formulations. Only a handful of studies have implemented FFF for non-biodegradable controlled-release drug-delivery devices (CDDD), mainly due to the poor printability of non-biodegradable, highly-elastic polymers such as ethylene-vinyl acetate (EVA) [2]. In this study, the 3D-printability of EVA9 (9% vinyl-acetate), EVA28 (28% vinyl-acetate) and a novel, non-biodegradable thermoplastic co-polyester (TPC) were compared for the first time. Furthermore, the drug-polymer interactions and drug-releasing properties of all polymers were investigated, using progesterone (P4) as a model drug. Finally, 3D-printed TPC-based urethra pessaries and implants were drug-loaded via impregnation and subjected to in-vitro drug-release studies. Compared to EVAs, the superior printability of TPC was a result of higher Young's modulus, lower melt viscosity and lower polymer crystallinity. The P4/polymer solubility followed the order EVA9<EVA28<TPC and was attributed to the lower TPC crystallinity (11.8±0.8%) compared to EVA9 (33.3±1.4%) and EVA28 (20.8±0.4%). The P4 diffusivity was mainly influenced by the density of the amorphous regions, following the order TPC≤EVA9<EVA28 (Fig. 1a). Based on these results, the expected drug-release was calculated and compared to the experimental data. It was found that, on top of the superior printability, TPC showed very similar drug-release rates to EVA28 and was, therefore, chosen for further studies. Drug-loading of 3D-printed samples via impregnation resulted in preferential P4 distribution within the outer specimen layers. For the implants, this was translated in pharmacologically relevant and relative constant drug fractions released over time. For the pessaries, high burst release was observed owing to the larger diffusional path required for the P4 solution to reach the core of the geometry and due to process-induced changes in the overall porosity of the pessary (Fig. 1b). Although the impregnation procedure requires sample geometry-specific adaptations, the results suggested that TPC can deliver high and tunable release-rates of P4 via a 3D-printed non-biodegradable CDDD.

Figure caption

Figure 1: (a) Progesterone solubility and diffusivity in TPC and EVA polymers, (b) progesterone release from 3D-printed pessaries and implants

References

1. Mohammed, A.; Elshaer, A.; Sareh, P.; Elsayed, M.; Hassanin, H. Additive Manufacturing Technologies for Drug Delivery Applications. *Int. J. Pharm.* **2020**, *580*, 119245.
2. Genina, N.; Holländer, J.; Jukarainen, H.; Mäkilä, E.; Salonen, J.; Sandler, N. Ethylene vinyl acetate (EVA) as a new drug carrier for 3D printed medical drug delivery devices. *Eur. J. Pharm. Sci.* **2016**, *90*, 53–63.

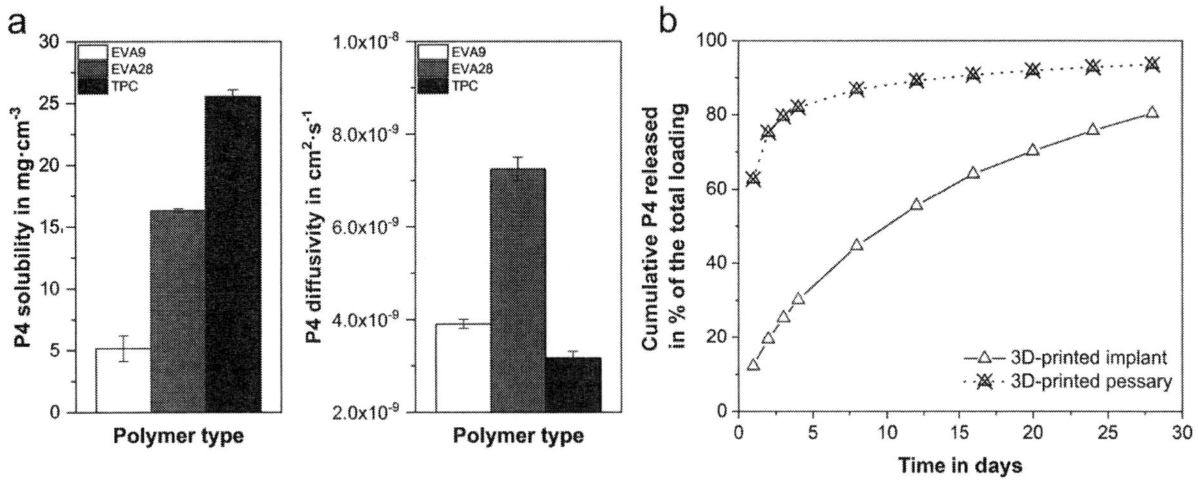
a
EVA9
EVA28
TPC
P4 solubility in mg·cm-3
Polymer type
P4 diffusivity in cm2·s-1
Polymer type
b
Cumulative P4 released in % of the total loading
Time in days
3D-printed implant
3D-printed pessary

624703 Investigation of Carrier Type and Blending Parameters on the Performance of a Dual-Combination DPI: A Comparative *in-Vitro-in-Silico* Study

Abstract:

Introduction

Powder drugs can be orally delivered to the lung using dry powder inhalers (DPIs). Two main types of DPI formulation exist namely: (i) carrier-free, consisting of soft active pharmaceutical ingredients (APIs) agglomerates and/or excipients with tailored properties and (ii) carrier-based, comprising a physical mixture of coarse carrier particles and APIs [1]. Most of the marketed formulations are carrier-based and offer the advantage of improved fine particle fraction (FPF) delivered to the patient, i.e. the quantity of an aerosolized API with particle size below 5 μm able to reach the lung.

It is important to understand the complex interplay between different parameters affecting the DPI performance, e.g. selection of carrier particles, API load, powder blending conditions, which are crucial to guarantee a stable dry adhesive mixture. However, one has also to consider how the formulation, and its respective performance will be impacted by the device and associated dispersion forces. To yield adhesive mixtures, the carrier needs to be properly blended with fine API particles by creating API-carrier interaction forces [2]. These subsequently need to be broken, also with the help of the DPI device, to detach the API from the carrier and allow its deposition in the deep lung.

Physiologically based pharmacokinetic modelling (PBPK) can help to better understand *in-vivo* DPI performance *in-silico*. By using PBPK, it is possible to gain, a *priori* understanding on how formulation parameters can affect the API behavior *in-vivo*. Therefore, this study aimed to use a comparative *in-vitro-in-silico* approach in which we investigated (i) how different carriers (α-lactose monohydrate (αLH) and mannitol (MAN) as potential alternative and (ii) blending parameters will impact the blend uniformity and aerosolization performance of formulations, as well as (iii) the impact of the device selection on the DPI performance. The data was used as an input in developed PBPK models and the potential impact on the predicted pharmacokinetic (PK) parameters studied.

Materials and Methods

Materials

Budesonide (BUD) and formoterol fumarate dihydrate (FF) were kindly provided by Chemo (Laboratorios Liconsa S.A., Spain) and Chiesi Farmaceutici S.p.A. (Italy), respectively. As carriers, the InhaLac® 230 (αLH, Meggle, Germany) and Parteck® M DPI (MAN, Merck, Germany) were used.

Blends preparation and characterization

The investigation of the impact of different carrier types and process parameters on the blending performance was performed by preparing αLH- and MAN-based blends of a model formulation (comprising 1 wt% of BUD and 0.02 wt% of FF) under different conditions. A total of 10 blends were prepared in a laboratory scale resonant acoustic mixer (LabRAM I ResonantAcoustic® Mixer, Resodyn™ Acoustic Mixers, USA) using a 236 mL vessel using the following blending parameters:

(i) 30 s at three acceleration levels (i.e. 30, 45 and 60 g) and (ii) 90 s at two acceleration levels (i.e. 30 and 60 g). The blend uniformity was evaluated by measuring the content of both APIs in 10 samples, taken within the powder bed at different positions. BUD and FF content was quantified using high performance liquid chromatography (HPLC). The blend uniformity was considered acceptable for RSD <10% and those blends were further analyzed for their aerodynamic performance.

Aerodynamic performance of blends

The impact of inhaler type on the aerosolization behavior was evaluated using three selected blends with acceptable blend uniformity. The assessment of aerodynamic performance was performed using the Next Generation Impactor (NGI; Copley Scientific, UK) comparing capsule-based (Cyclohaler®) and reservoir-based (Novolizer®) inhalers. All experiments were carried out in triplicates and at a flow of 60 L/min. For Cyclohaler®, formulations were loaded into 20 hydroxypropyl methylcellulose (HPMC) capsules size 3, whereas in the case of Novolizer®, 40 actuations sufficed.

PBPK model development and predictions

Based on the reported *in-vivo* data following inhalation of Symbicort Turbuhaler®[3], the PBPK models for FF and BUD were developed using MATLAB® (R2021, MathWorks, USA). The models were developed based on the previously published model by Boger et al [4]. To predict the deposition patterns of FF and BUD [5], their aerodynamic size distributions and spirometry profiles were used as inputs in Multiple-Path Particle Dosimetry (MPPD; Applied Research Associates, Inc.).

Results and Discussion

Powders used for inhalation can either contain a single drug or a combination of two or more APIs. In this study, we selected a combination product of BUD and FF in the same ratio as in commercial products (i.e. Symbicort). Evaluation of blend uniformity was performed following each blending experiment. For αLH containing blends, a blending time of 30 s in combination with an acceleration of 60 g resulted in satisfactory homogeneity (Figure 1). For MAN based blends instead, homogeneity was achieved following 90 s of blending at both acceleration rates (i.e. 30 and 60 g). Interestingly, different blending trends of BUD and FF were observed when comparing αLH and MAN as carriers. When using αLH as carrier, the homogeneity of FF was improved by higher blending times and accelerations, contrary to BUD where de-mixing was observed. For mixtures containing MAN, a higher mixing time was necessary to evenly distribute cohesive FF particles, independently of the acceleration level, while BUD homogeneity was dependent on both time and the acceleration levels. The observed effect can be attributed to a higher surface area of MAN, hence for the same mass of API, longer time was needed to evenly distribute it over the bigger carrier particles. Additionally, the mixing behavior and efficiency might be impacted by difference in loading of the two API.

Ideally, the DPI device will allow the patients to generate the preferred airflow e.g. 60 L/min and will deliver constant and high FPF to the patients. Generally, two main types of DPI devices are single-dose (e.g. capsule-based) and multi-dose (e.g. reservoir). We selected Cyclohaler® (capsule) and Novolizer® (reservoir) for our study. Three blends with satisfactory homogeneity (i.e., one αLH mixture obtained after blending at 60 g for 30 s and two MAN mixtures prepared at 30 and 60 g for 90 s) were further evaluated in terms of their aerodynamic performance. Interestingly, for all blends a higher FPF was seen when the reservoir-based inhaler was used, i.e. the FPF values were 30% for the αLH-based blend and 40% for the MAN ones. For capsule-based systems, the FPF was in fact around 23-24% for both αLH- and MAN-based blends.

The mass mean aerodynamic diameter (MMAD) of the delivered particles was around 1 μm for MAN containing blends compared to the αLH ones, where it was around 2 μm, for both inhalers. These observed differences in aerodynamic performance were further evaluated by use of PBPK models. It is expected that different deposition patterns could potentially impact the *in-vivo* performance of the combination formulation. The models showed that distinct changes in API deposition can occur in case when different carriers are used, resulting in different bioavailability of the two drugs. Additionally, the selection of the device showed to be a critical parameter that impacting the performance of the formulations.

Conclusion

The present study highlights the importance of a proper selection of carrier type, blending process parameters and inhaler device for the design of an efficient DPI combination product. Compared to αLH, the selection of MAN as carrier in combination with a higher blending time and a reservoir based type of inhaler resulted in the improved aerodynamic performance of BUD and FF. Utilization of PBPK models demonstrated how a combined *in-vitro-in-silico* approach can be used to understand a priori the impact of different formulation approaches on the *in-vivo* behavior of DPIs.

References

[1] D. Traini, *Inhalation Drug Delivery*. 2013.

[2] W. Kaialy, "On the effects of blending, physicochemical properties, and their interactions on the performance of carrier-based dry powders for inhalation — A review," *Adv. Colloid Interface Sci.*, vol. 235, pp. 70–89, 2016.

[3] S. Lähelmä *et al.*, "Equivalent lung dose and systemic exposure of budesonide/formoterol combination via easyhaler and turbuhaler," *J. Aerosol Med. Pulm. Drug Deliv.*, vol. 28, no. 6, pp. 462–473, 2015.

[4] E. Boger and O. Wigström, "A Partial Differential Equation Approach to Inhalation Physiologically Based Pharmacokinetic Modeling," *CPT Pharmacometrics Syst. Pharmacol.*, vol. 7, no. 10, pp. 638–646, 2018.

[5] Á. Farkas *et al.*, "Numerical simulation of emitted particle characteristics and airway deposition distribution of Symbicort® Turbuhaler® dry powder fixed combination aerosol drug," *Eur. J. Pharm. Sci.*, vol. 93, pp. 371–379, 2016.

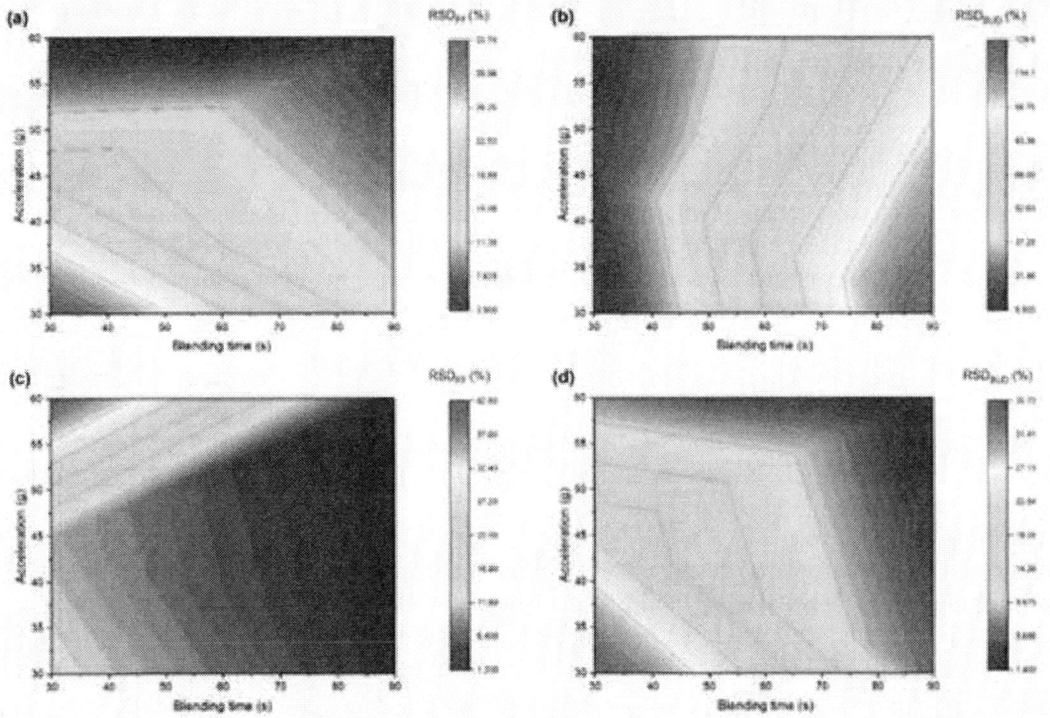

***Figure 1**: Effect of blending process parameters on the blend uniformity (RSD) of the prepared adhesive mixtures: (a) FF and (b) BUD in αLH-based mixtures; (c) FF and (d) BUD in MAN-based mixtures*

622604 Nanoemulsion-Loaded Capsules for Controlled Delivery of Lipophilic Active Ingredients

Abstract:

Nanoemulsions have gained considerable attention in recent years because of the increasing need for delivering lipophilic active ingredients that are ubiquitous and indispensable in pharmaceutical, cosmetic, and food applications. However, lipophilic active ingredients are hydrophobic in nature, which limits their bioavailability and absorption efficiency. Nanoemulsions featuring their uniform nanodroplets can not only address the bioavailability issue but also provide high solubilization capacity and enhanced stability. Building on these ideal properties of nanoemulsions, encapsulation technologies have been developed to formulate nanoemulsions into dosage forms for incorporating more benefits: accurate administration, controlled delivery, and enhanced stability. Nevertheless, nanoemulsion encapsulation techniques are limited to monolithic dosage forms, and the lack of versatile carrier platforms for nanoemulsions hinders advanced control over their release behavior.

In this work, a method is developed to encapsulate nanoemulsions in alginate capsules for the controlled delivery of lipophilic active ingredients. Functional nanoemulsions loaded with active ingredients and calcium ions are first prepared, followed by encapsulation inside alginate shells. The intrinsically high viscosity of the nanoemulsions ensures the formation of spherical capsules and high encapsulation efficiency during the synthesis. A facile approach is developed to measure the nanoemulsion release profile from capsules through UV-Vis measurement without an additional extraction step. A quantitative analysis of the release profiles shows that the capsule systems possess a tunable, delayed-burst release. The proposed encapsulation methodology can be further generalized to other functional nanoemulsions with various active ingredients, oil phases, nanodroplet sizes, and chemically crosslinked inner hydrogel cores. Overall, the capsule systems provide promising platforms for various functional nanoemulsion formulations.

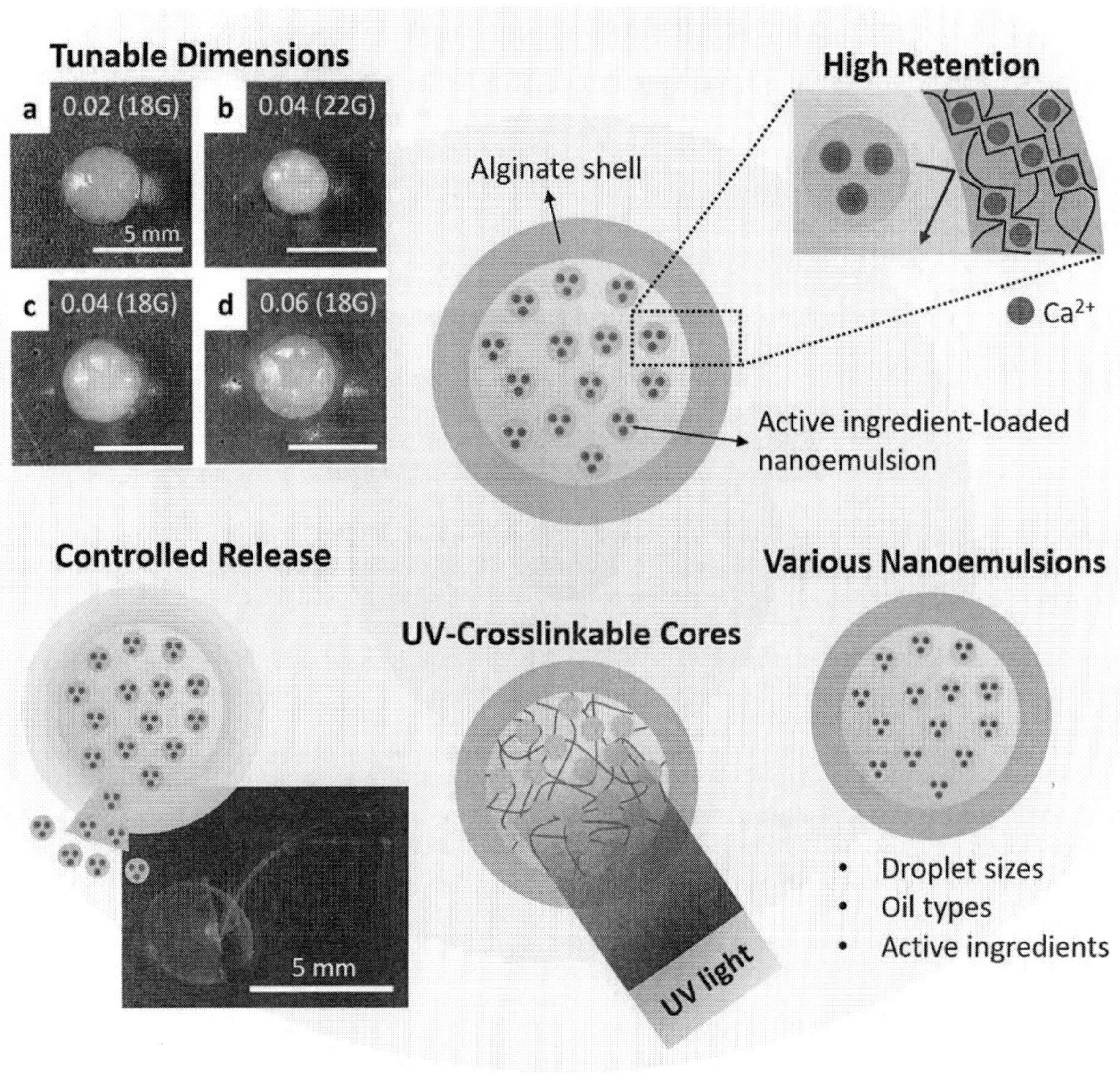
Tunable Dimensions
a
0.02 (18G)
5 mm
b
0.04 (22G)
c
0.04 (18G)
d
0.06 (18G)
Alginate shell
High Retention
Ca2+
Active ingredient-loaded
nanoemulsion
Controlled Release
5 mm
UV-Crosslinkable Cores
UV light
Various Nanoemulsions
Droplet sizes
Oil types
Active ingredients

627945 Transport- and Reaction-Modeling of Nanocarriers for Cancer Therapeutics Via Experimental and *in-Silico* approaches

Abstract:

Introduction: Solid tumors account for 80-90% of all reported cancer cases. Various therapies leveraging different properties exhibited by tumors have been experimented with over the last few years. At the same time, there have been multiple mathematical models developed to model the response of cancer cells and of tumors to therapies. Yet, we realize that there is a disconnect between the two communities: sophisticated nanoparticles are reported usually without dosimetric analysis, which could in principle enable extrapolation of the reported results to other tumors and/or types of therapeutics. The experimental community has yet to fully embrace the functionality of the modeling, as it is rarely "personalizable" and, thus, difficult to exploit. It is abundantly clear that cross-fertilization between the experimental and the *in-silico* research approaches to solid tumor drug delivery carrier design should be particularly valuable for fast turnaround results. A user friendly, personalizable tool, UMA (User-friendly Modular Algorithm) of this kind, which could (a) help estimate/evaluate the limitations and abilities of drug delivery carriers currently under development without the need for extensively time-consuming experiments and (b) allow for designs of hypothetical carriers exploring future materials, is a much needed development as such a tool does not currently exist. We are interested in the interplay of different (and novel) carrier design options, different transport and binding steps –many of them pH profile dependent- and different spheroid types (used as *in vitro* surrogates of the avascular tumor regions).

Materials and Methods: To characterize multicellular spheroids comprising the HER2-positive and triple negative breast cancer cell lines BT-474 and MDA-MB-231, respectively, the interstitial pH_e profiles of different spheroid sizes were evaluated using a membrane impermeant (SNARF-4F) fluorescent pH indicator. Liposomes were used as ideal analog simulators of other drug carriers and were designed to have different behavioral characteristics such as binding specificity, drug release kinetics, cell internalization etc. and were utilized to help validate the *in-silico* model. Liposomes possessing combinations of these properties were formed using the thin film hydration method. They were further characterized by evaluating the aforementioned properties. Doxorubicin (DXR) and cisplatin (CDDP) were chosen to be evaluated as the delivered therapeutic agents, and the dose responses to these agents were evaluated on the same cell lines. Temporal microdistributions of liposomal carriers and of their therapeutic cargo were also evaluated in spheroids.

Results and Discussion: The efficacy of a drug-carrier system is determined as the combined result of different processes, namely, transport, binding, drug release and drug availability at the active site(s). Experiments were performed to isolate each process - when possible - and to evaluate the necessary and required parameters, characterizing these processes. Parameters obtained by fitting the processes (including but not limited to binding, internalization, trafficking, release, killing etc.) to rate laws were used to experimentally inform the mathematical model. Since the focus of this study is on established tumors which develop acidic interstitial regions, in several of the processes mentioned above, parameters were found to be a function of the local pH which is in turn a function of the location within the spheroid (i.e. the radial distance).

Conclusion: The developed tool is intended to serve as a guide to reduce the number of iterations in designing novel carrier systems and in addressing specific challenges in therapy by taking into account the intratumoral and intertumoral heterogeneities. The developed tool will go beyond simply validation and help multiple labs evaluate multiple carrier systems and make informative and quantitative decisions about their therapeutic potential.

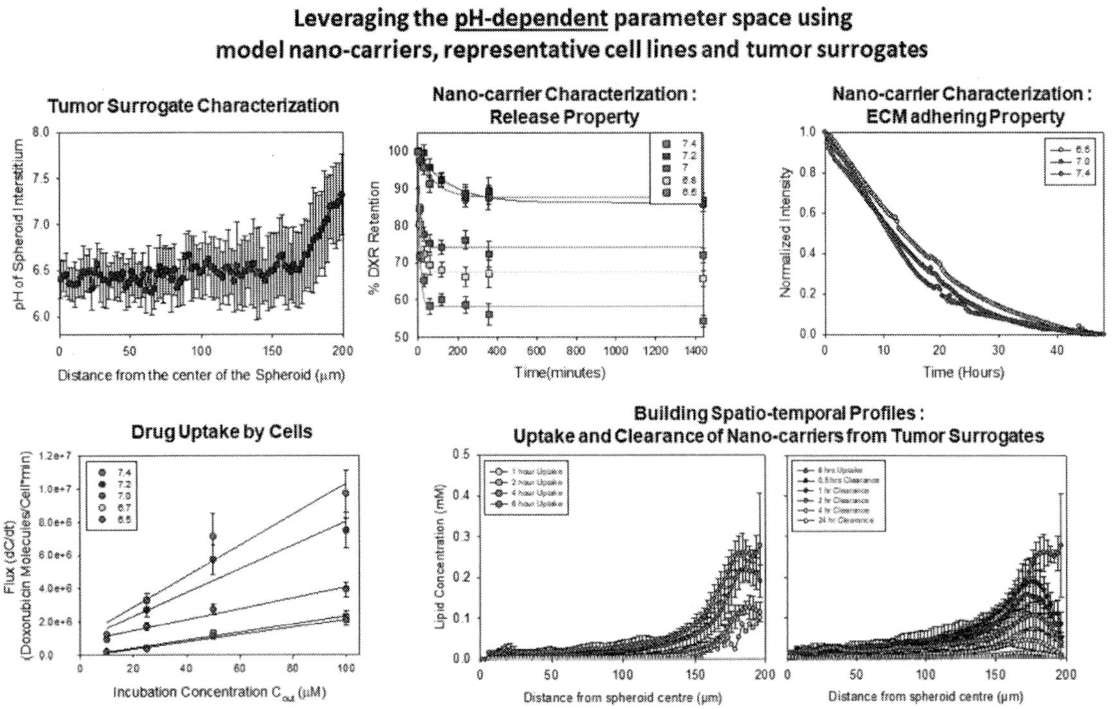
Leveraging the pH-dependent parameter space using model nano-carriers, representative cell lines and tumor surrogates
Tumor Surrogate Characterization
pH of Spheroid Interstitium
Distance from the center of the Spheroid (μm)
Nano-carrier Characterization : Release Property
% DXR Retention
Time(minutes)
Nano-carrier Characterization : ECM adhering Property
Normalized Intensity
Time (Hours)
Drug Uptake by Cells
Flux (dC/dt) (Doxorubicin Molecules/Cell*min)
Incubation Concentration C_out (μM)
Building Spatio-temporal Profiles : Uptake and Clearance of Nano-carriers from Tumor Surrogates
Lipid Concentration (mM)
Distance from spheroid centre (μm)
Distance from spheroid centre (μm)

629300 Spray-Dried Lipid-Microparticles for Inhalation: Correlation between Solid State and Processability

Abstract:

Motivation and scope

Spray-drying is a versatile process in which a solution containing solid components is atomized into a hot gas promoting rapid evaporative mass transfer. Targeted particle attributes are provided by the interplay of material science and process parameters. Inhalable particles, composed of active pharmaceutical ingredients and specialized excipients, can be readily yielded. However, technical hurdles are commonly experienced when lipid-based excipients (LBE) are spray-dried. Sticky products, extensive deposition on the spray-dryer walls, or simply inability to yield a solid product have been associated to the low melting point of LBE. Comprehensive knowledge on the correlation between lipid solid state and process boundaries can lead to enhanced processability. In this work, several LBE possessing high melting points were screened in combination with ibuprofen (IBU) as a model drug. Particle attributes were envisaged to provide systemic delivery of IBU after pulmonary administration. The following LBE were employed: behenoyl polyoxyl-8 glycerides (BPG) available as Compritol® HD5 ATO, tripalmitin (PPP) available as Dynasan® 116, and triacylglycerol ester of behenic acid (PG3C22p) available as Witepsol® PMF 123.

Methods

Prior to spray-drying the solid state interactions of LBE and IBU were investigated via binary phase diagrams. Alteration of the melting points allowed the identification of temperature boundaries of spray-drying and selection of API-loading. Subsequently, solutions of LBE:IBU at 1.5%w/w solid content in tetrahydrofuran (THF) were prepared. Viscosity, surface tension and density were determined. The solutions were spray-dried. Mass and heat transfer rates and droplet drying kinetics were set constant to the most extent for all LBE:IBU by controlling the outlet temperature (T_{out}) and initial droplet size. Spray-drying was carried out at inlet temperature (T_{in}) of 71°C, using nitrogen as stagnant drying gas at 0.3m^3/h. Atomization was carried out at feed rate of 3.2 g/min through a 0.2mm bi-fluid nozzle and gas to liquid mass flow ratio of 1.45. The process yield was obtained and resulting particles were characterized. Lastly, the thermal behavior and crystallization of LBE was assessed via differential scanning calorimetry (DSC) and small and wide x-ray scattering (SWAXS) and correlated to the processability.

Results

The phase diagrams LBE:IBU evidenced depression of melting points due to eutectic interaction. Liquid and solid regions were identified. The eutectic composition, having the least solid-liquid transitions at the eutectic temperature (T_{eu}), was selected for spray-drying. Since it is generally accepted that during spray-drying, particles experience most of their residence time at T_{out}, a rational criterion to guarantee processability was to achieve T_{out} below $T_{eu.}$ This would assure the existence of a unique solid region decreasing the risk of softening and melting at the outlet. T_{eu} was found to be T_{BPG}=45.7, $T_{PG3C22p}$=60.5 and T_{PPP}=60.1°C at LBE:IBU ratio of 70:30.

Viscosity and surface tension of all LBE:IBU solutions were not significantly different, therefore atomization led to nearly equal initial droplet size. Semi-empirical estimation of the droplet Sauter mean diameter (DSMD) based on Weber and Onhesorge numbers yielded values between 8.931 and 9.012 μm. The combination of process parameters led to T_{out} between 32-35°C, which guaranteed the existence of solid region for all LBE:IBU.

Despite having equal initial droplet size, analogous mass-heat transfer and T_{out} below T_{eu}, the resulting outcome from each LBE:IBU was markedly different. Agglomeration and low yield were depicted by BPG (y=28.6%) with sauter mean diameter (SMD=107.7μm) notably larger than DSMD. PG3C22p and PPP, although having the same T_{eu} showed yields of 76.2 and 10.6%, respectively. PPP displayed agglomeration (SMD=19.19μm), bimodal particle size distribution and general impairments, whereas PG3C22p yielded uniform non-agglomerated particles (SMD=1.973μm).

It is reasonable to deduce that a minimum difference between T_{eu} and T_{out} should be provided. Produced particles stayed in the collection vessel at T_{out} until the feedstock was completely atomized. Meanwhile they followed the characteristic rotational movement of the gas in the vessel (caused by the cyclone) which potentially created friction among particles and sections of higher temperature. The closeness of the T_{out} to T_{eu} increases the risk of softening; as seen for BPG.

However, this should not be the only consideration for lipid-based spray-drying. During spray-drying, the droplet temperature reaches the saturation temperature (T_s), that is the wet bulb temperature of a nitrogen-solvent vapor (THF in this study) mixture. T_s is usually around 10-15°C below the T_{out}. When the particle is crystallized, its temperature is promptly increased to the dry bulb temperature of the surrounding gas. A particle recently crystallized near the atomization centerline will directly exit the drying chamber, experiencing the shortest residence time and a highest temperature equal to T_{out}. Whereas particles that get entrained within a recirculating eddy will experience longer residence time and exposition to the high T_{in} without the protective cooling effect of THF-vapor. Reaching T_{in}=71°C leads to instant melting for all LBE:IBU. As the time scale of the process is very short, the temperature of the recently melted particle promptly decreases as it moves towards the outlet. The melted particle experiences a temperature profile from T_{in} to T_{out} leading to recrystallization from the melt.

PG3C22p melted particles recrystallized almost immediately, since its crystallization temperature (T_c=57.4°C) is very close to the T_{eu}. This is due to the characteristic absence of polymorphism and phase separation of polyglycerol esters of fatty acids (PGFAs). Whereas PPP recrystallizes from the melt at T_c=31.4°C into a different polymorph (α-form). Besides the simply relation to T_{out}, presented above, BPG melted in two phases and recrystallized into five, with the lowest T_c at 22.8°C. The impairments of spray-drying of BPG and PPP can be attributed to the existence of molten fractions, not being able to recrystallize at the surrounding drying gas temperature due to polymorphism or multi-phases.

PG3C22p, in combination with IBU, was effectively spray-dried. Lipid micro-particles of volume mean diameter of 6.586±0.543μm, density of 0.389±0.007 g/cm^3 and corrugated surface were produced. The application of the lipid-microparticles as carrier-free dry powder for inhalation was tested using Aerolizer coupled with the next generation impactor. High emitted fraction (>90%), mass median aerodynamic diameter of 3.568±0.113μm and fine particle fraction of 45.6±1.6% were achieved. Although a LBE was employed, high yield of microparticles was obtained by a simple spray-drying process. No device modifications or material-aids were required.

Conclusion

The absence of processability of LBE via spray-drying has been associated to their low melting points. Sufficiently low T_{out} are typically recommended to avoid product loss. However, it was demonstrated that, LBE having the same melting temperature and spray-dried at low T_{out}, result in entirely different yields and particle attributes. The complex crystallization pathways of LBE strongly impact the process performance even at sufficiently low T_{out}. Crystallization into polymorphic forms or multi-phases negatively impact the process yield associated to the formation of liquid fractions of long residence time in the spray-drying chamber. The application of high melting LBE undergoing minimum solid state transitions, such as PG3C22p, can broaden the operational window of spray-drying leading to further possibilities of particle engineering.

Acknowledgment: IOI Oleo GmbH, the Austrian Funding Agency (FFG) and the Doctoral Academy NanoGraz, University of Graz.

630131 Acid-Responsive Colloidal Drug Aggregates Escape from the Endolysosomal Pathway

Abstract:

Problem

The discovery and development of new drugs is immensely costly, with the average cost of bringing a drug to market cited at more than $1 billion. Early in this process, thousands of candidates are tested against the desired target. Here it was discovered that the poor solubility of many candidate drugs causes them to form colloidal aggregates above a critical aggregation concentration. This behavior leads to both false positives in drug screening assays that rely on enzyme inhibition and false negatives in cytotoxicity assays. Such misleading results waste resources and so colloidal aggregation has traditionally been seen as a nuisance during drug development.

Despite these in vitro challenges, colloidal drug aggregates could be useful as vehicles for drug delivery in vivo. Drugs that form colloidal aggregates often require solubilizing excipients due to their poor solubility; however, these additives frequently cause dose-limiting side effects. Thus, formulations that can safely deliver high concentrations of active substance with minimal excipients are desperately needed. Colloidal drug aggregates are uniquely positioned to fill this gap due to their drug-rich nature. Co-assembly with dye molecules and/or polymeric surfactants has stabilized a series of drugs as nano-sized colloidal drug aggregates with drug loadings as high as 95%.

However, upon endocytosis, these colloids do not readily escape the endolysosomal pathway, resulting in sub-optimal efficacy. These developments have led to the new challenge of designing formulations that can escape the endosomes into the cytosol, resulting in optimal potency.

We hypothesized that colloid-forming drugs that are ionized in the acidic environment of the endolysosomal pathway could accomplish this objective by disrupting the integrity of the phospholipid membrane that separates the endolysosomes from the cytosol. This approach is supported by literature demonstrating that ionizable lipids, polymers, and small molecule drugs can cause membrane disruption through membrane fusion, osmotic pressure effects, or changes in lipid homeostasis.

To demonstrate this concept, we have previously formulated stable co-colloidal drug aggregates composed of two cancer drugs, fulvestrant and lapatinib, and demonstrated that lapatinib release can be triggered by acidification to endosomal pH. Transferrin-coated colloids facilitated uptake by cancer cells, and the acid-responsive ionization behavior of lapatinib in the colloids resulted in its release into the acidic endo-lysosomal pathway of cancer cells, leading to enhanced cytotoxicity.

Despite the success of this approach, it is limited to drugs that natively respond to changes in pH. In this work, we demonstrate a strategy to overcome this limitation by introducing pH-responsive functionality to a non-ionizable drug.

Methods

Fulvestrant, a breast cancer drug that forms colloidal aggregates that are not acid-responsive, was modified with different ionizable functional groups in a way that would preserve its bioactivity. Previous research has indicated that the estradiol core of fulvestrant is critical for its antiestrogen activity whereas the tail group tolerates some modifications. Therefore, the ionizable groups were attached onto the sulfur atom present on the tail of fulvestrant.

Next, the ionization behavior of these fulvestrant analogs at was evaluated as a function of pH, because ionization in the reduced pH of the endolysosomal pathway (pH ~4.5-6.5) is expected to be necessary for endosomal escape of colloidal drug aggregates. The pKa, a property that describes a molecule's ionization state as a function of pH, was measured for each of the fulvestrant analogs using a fluorescence assay. This assay, which is commonly used to characterize ionizable lipids, determines pKa by measuring the fluorescence of added 6-(p-toluidino)-2-naphthalenesulfonic acid (TNS) over a range of pH values.

Fulvestrant analogues were then tested for formation colloidal drug aggregates. Solutions of each analog in dimethyl sulfoxide were optionally mixed with solutions of excipients and then a large excess of phosphate-buffered saline solution (PBS) was added. The addition of PBS causes drug phase separation into nano-sized aggregates with the stabilizing excipients coating their surfaces. The hydrodynamic diameter and polydispersity of colloidal drug aggregates were characterized by dynamic light scattering.

A phenotypic assay for endosomal escape was developed. Conclusively quantifying endosomal escape by particle tracking is notoriously difficult; therefore a fluorescent reporter for endosomal escape was chosen. This approach involved generating a cell line that constitutively expresses a fluorescent galectin 8 fusion protein. Galectin 8 is normally spread throughout the cell cytosol, resulting in dim and diffuse fluorescence. When an endosome disruption event occurs, galectin 8 can access and bind to carbohydrates that are present on the inner side of the endosomal membrane, causing a high local galectin 8 concentration and bright fluorescence. An imaging pipeline for identifying and counting the resulting bright puncta, thereby objectively measuring endosomal escape, was then developed.

Finally, the ability of the fulvestrant analogs and their colloids to cause endosome disruption was tested. To test the endosome disruption activity of each analog in isolation, fluorescent galectin 8 reporter cells were treated with a low concentration of each analog for 24 h. Then, the ability of fulvestrant analog colloidal drug aggregates to escape from endosomes was measured. Colloids were formulated as detailed above with the addition of a hydrophobic dye to enable their visualization inside cells. Then, the reporter cells were treated with these formulations for 3 h and subsequently imaged to measure the extent of colloid uptake and endosome disruption.

Results

Using a 5-step synthetic sequence, we successfully produced thirteen different fulvestrant analogs each containing a different ionizable functional group on its tail. We tested whether these analogs are still bioactive and found that they remained active against MCF7 cells with growth IC50 values below 10 nM. The potency of most of the analogs was not significantly different from the potency of the parent drug (IC50 close to 2 nM), validating our judicious selection of the site of modification.

Using the TNS method, we found that the fulvestrant analogs spanned a range of pKa values from 5 to 8. We found that the structure of the functional group was correlated to the pKa, with the inclusion of electron-withdrawing groups reducing the pKa. Importantly, the pKa range of the chosen functional groups would allow their protonation in the pH range of the endolysosomal pathway (pH ~4.5-6.5).

We found that all the fulvestrant analogs formed colloidal drug aggregates above 1 μM, which is consistent with the behavior of the parent compound. Although colloids without any excipients formed large, polydisperse aggregates in PBS, we found that 100 nm colloids with polydispersity indices less than 0.2 could be consistently formulated when small amounts of stabilizing excipients were added before formation of the colloids in PBS. The drug loading of these formulations exceeded 50% which is consistent with our goal to minimize excipient burden.

After infecting the parent cell line with lentivirus encoding a fluorescent mCherry-galectin 8 fusion protein and selecting for transfected cells with puromycin, we isolated a population of cells that consistently express the fluorescent fusion protein over multiple passages. We verified that the fluorescence was spread throughout the cytosol of untreated cells but distributed in a punctate pattern in cells whose endosomes and lysosomes were disrupted by siramesine, a known lysosomotropic drug.

We were excited to find that many of the fulvestrant analogs disrupted endosomes. Measuring the extent of endosome disruption by the fulvestrant analogs, we found that there was a striking correlation between pKa and endolysosomal disruption, with analogues having a pKa between 7 and 8 disrupting the greatest number of endosomes per cell. This result was consistent with the known endosome disruptor siramesine which we found to have a pKa of 7.4.

When treating cells with colloidal drug aggregates containing the ionizable fulvestrant analogs, we observed robust uptake and endosomal escape. The number of disrupted endosomes was reduced when colloid endocytosis was blocked, confirming that the observed endosome disruption was caused by endocytosed colloids. Colocalization between colloids and disrupted endosomes further demonstrated that these colloids can disrupt the integrity of the endosomes that contain them. These results suggest a fusion or leakage mechanism of release where the colloid does not escape the endosome intact but may still release its cargo into the cytosol.

Implications

This work demonstrates a widely applicable strategy for achieving endosomal escape of colloidal drug aggregates. Beyond this goal, the proposed work has broad implications in drug development and nanomedicine. Current dogma focuses on optimizing drugs for systemic delivery as individual small molecules, but these formulations are hindered by limited solubility and toxic excipients. By delivering drugs as pH-sensitive, stable colloids, the use of excipients can be reduced, and dosing limitations can be overcome. This new way of looking at colloidal drug aggregates addresses an issue that continues to plague drug development for multiple disease areas. With our platform technology, we aim to change drug design and delivery.

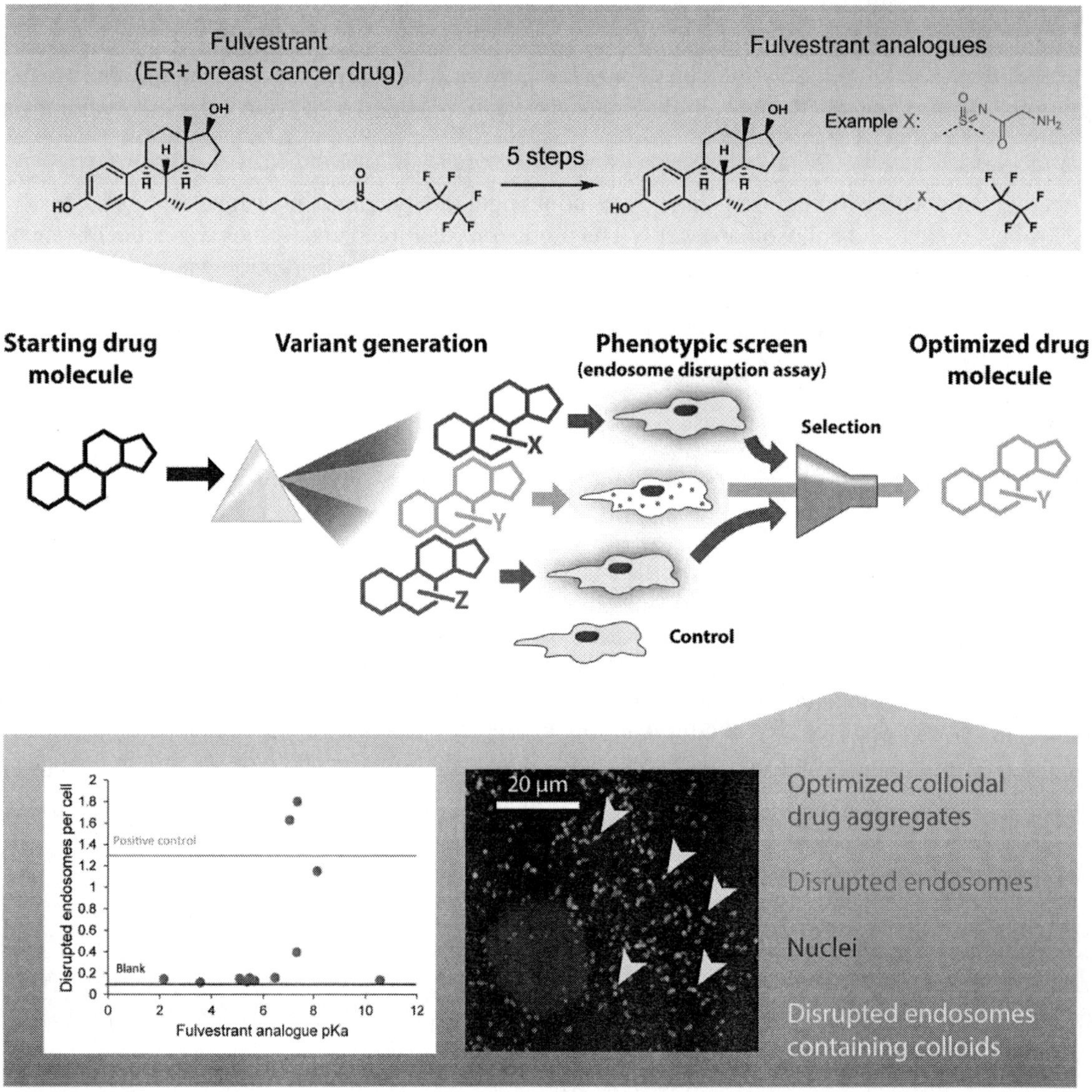
Fulvestrant
(ER+ breast cancer drug)
Fulvestrant analogues
Example X:
5 steps
Starting drug molecule
Variant generation
Phenotypic screen
(endosome disruption assay)
Optimized drug molecule
Selection
Control
Disrupted endosomes per cell
Positive control
Blank
Fulvestrant analogue pKa
20 µm
Optimized colloidal drug aggregates
Disrupted endosomes
Nuclei
Disrupted endosomes containing colloids

623346 Process Modeling and Control of Digital Biopharmaceutical Manufacturing

Abstract:

Recent trends in biopharmaceutical manufacturing provide opportunities for process systems engineering to make major advances in biomanufacturing: (1) process analytical technology (PAT) is providing on-line measurements of critical quality attributes (CQAs) for constructing first-principles and data-based models of each unit operation and enable advanced control, (2) a transition from batch to continuous operation provides process control needs to handle the propagation of impurities and other disturbances caused by tight integration of unit operations, and (3) the invention of new designs for downstream processes is creating new processes to control [1].

These trends ultimately lead biomanufacturing to digital manufacturing, which is an integrated approach to manufacturing centered around a computer system. Digital manufacturing is enhanced by using modern system engineering tools. Modeling and simulation provide increased mechanistic fidelity, better empirical approaches, and automated decision making. Process optimization enables drug- and patient- specific manufacturing and reduces experimental costs. Process control deals with process variations and enables fully automated systems to minimize operator error [2].

This presentation describes how modern system engineering tools are applied for digitalization of biomanufacturing systems through process modeling and control. The presentation describes mathematical models and optimal control methods for multiple bioreactor configurations including microbioreactor systems [3] and stirred-tank bioreactors [4]. The presentation describes laboratory unit operation systems constructed for downstream process including protein crystallization [5] and continuous viral inactivation. These systems are optimally designed and controlled based on first-principles models.

References:

[1] M. S. Hong, K. A. Severson, M. Jiang, A. E. Lu, J. C. Love, and R. D. Braatz, Challenges and opportunities in biopharmaceutical manufacturing control, *Computers & Chemical Engineering*, 110:106-114, 2018.

[2] M. S. Hong, W. Sun, A. E. Lu, and R. D. Braatz. Process analytical technology and digital biomanufacturing of monoclonal antibodies. *American Pharmaceutical Review*, 23(6):122-125, 2020.

[3] M. S. Hong and R. D. Braatz. Mechanistic modeling and parameter-adaptive nonlinear model predictive control of a microbioreactor. *Computers & Chemical Engineering*, 147:107255, 2021.

[4] M. S. Hong, M. L. Velez-Suberbie, A. J. Maloney, A. Biedermann, K. R. Love, J. C. Love, T. K. Mukhopadhyay, and R. D. Braatz. Macroscopic modeling of bioreactors for recombinant protein producing Pichia pastoris in defined medium. *Biotechnology & Bioengineering*, 118(3):1199-1212, 2021.

[5] M. S. Hong, K. Kaur, N. Sawant, S. B. Joshi, D. B. Volkin, and R. D. Braatz. Crystallization of a non-replicating rotavirus vaccine candidate. *Biotechnology & Bioengineering*, 118(4):1750-1756, 2021.

623979 The Use of Hybrid Modeling Schemes in the Development of a Probabilistic Condition Monitoring System for a Continuous Drug Product Manufacturing Process

Abstract:

Industry 4.0 concepts have the potential to address problems in manufacturing operations and development. Through the seamless integration of separate manufacturing systems, Industry 4.0 results in the development of a "smart factory" that is lean, agile, and flexible.(Steinwandter, Borchert, and Herwig 2019; Barenji et al. 2019) Such a system features a centralized data collection system, paving the way for a more holistic way to analyze data. The application of machine learning models have played a significant role in taking advantage of this trend, allowing industries to effectively derive information and knowledge from massive amounts of data.(Qin 2014)

While the same can be said about Pharma 4.0 (Industry 4.0 for the pharmaceutical industry), unique challenges in pharmaceutical manufacturing, especially in drug development, make the adoption of these trends difficult. The drug development period is regarded as the critical factor in the success of a drug product. (Sugiyama et al. 2019) It is also during this period that data scarcity becomes an issue. During this stage, it would typically be very expensive to produce the relevant drug substance and its derivatives in large amounts, limiting the amount of material that can be used for data generation. Concurrently, models for advanced process control and condition monitoring need to be developed and proven to be effective in quality control and assurance. This is especially true for Continuous Manufacturing Systems, where the timely detection of faults and application of intervention measures are critical for its practicability.(Schenkendorf 2016)

This presentation features a case-study where a probabilistic condition monitoring model for Dry Granulation was developed under scarce data conditions. A limited amount of data was used to estimate the parameters of a flowsheet model, which serves as a digital twin for the process. Using this virtual representation of the process, data is generated for developing a probabilistic condition monitoring model, which can be more data-intensive.

Digital twins have been proven to simulate particulate systems effectively. The following figure demonstrates the close match between the actual experimental data from an impact pin mill, and predictions from its digital twin, which was developed using a multiscale modeling approach. (Wang et al. 2021)

Figure 1. Particle Size Measurements from an Impact Pin Mill versus the Predictions of its Digital Twin

Models like these are suitable for process predictions since they are based on first principles approaches like Population Balances, (Chaudhury et al. 2016) and fitted with actual data from operations. For the dry granulation method of tablet manufacturing, estimating the particle size distribution is of particular interest since it is directly related to the flowability and compressibility of the particles. These two parameters are important because they are Critical Material Attributes (CMA)(Maguire and Peng 2015) for maintaining product quality in the Tablet Press (TP). They are also considered the critical-to-quality attributes of the Roller Compactor (RC) by default, since the RC directly precedes the TP along the dry granulation line.

The identification of Quality-by-Design (QBD) parameters such as (Intermediate)-Critical-Quality Attributes (CQA) and Critical Process Parameters (CPP) proved to be an advantage as it caters to the utilization of Model-based Machine Learning Methods(Bishop 2013). These methods utilize graphical models based on process understanding to develop bespoke machine learning algorithms. An example of this model is shown in Figure 2, which represents the relationship between the condition of the process (variable A), the condition of the two sensors (variables B and E), and the readings from the two sensors. These graphs are called directed acyclic(Koller and Friedman 2009) graphs, and the nodes in these graphs are random variables with a certain probability distribution.

Figure 2. Directed Acyclic Graph (Bayesian Network) of a Process with Two Sensors

This graphical model can be converted to a factor graph, which contains nodes between connected random variables to represent their mathematical relationship. Factor graphs can be easily programmed using a probabilistic programming framework like Infer.NET(Minka et al., 2018), where machine learning is implemented using a Bayesian approach. The random variables of interest (i.e. the parameters that need to be predicted) have designated distributions called the prior, and these are updated using data observed from other variables (i.e. the input variables, a.k.a. features). Using message-passing algorithms, the distribution of unobserved random variables of interest are "inferred", producing a posterior distribution which updates the prior.(Bishop 2013) This is a different paradigm from traditional machine learning where parameters are assigned point estimates, and the values of these point estimates are optimized by minimizing a loss function that reflects the difference between predicted and observed values.(Bishop 2013; Martin 2018) An example of a machine learning algorithm developed using probabilistic programming is shown in Figure 3, where process information from the Roller Compactor was used to predict the faults.

Figure 4. Probabilistic Graphical Model of a Roller Compactor Fault Detection Tool

The key advantage of the probabilistic approach is the natural quantification of uncertainty. Each prediction made by the model has an associated probability that reflects the certainty of that prediction. From an application perspective, this gives the operator the power to make informed decisions on whether to trust the predictions made by the system. From a model development perspective, this facilitates model comparison as well as guide the modeler to ascertain if the training data is sufficient for creating a reliable tool for fault detection. Designing experiments to produce training data can now become decisions based on quantifiable properties, placing process engineers in a position to create a better business case for justifying the cost of acquiring (or not acquiring) additional training data.

References

Barenji, Reza Vatankhah, Yagmur Akdag, Barbaros Yet, and Levent Oner. 2019. "Cyber-Physical-Based PAT (CPbPAT) Framework for Pharma 4.0." *International Journal of Pharmaceutics* 567 (August): 118445.

Bishop, Christopher M. 2013. "Model-Based Machine Learning." *Philosophical Transactions. Series A, Mathematical, Physical, and Engineering Sciences* 371 (1984): 20120222.

Chaudhury, Anwesha, Maitraye Sen, Dana Barrasso, and Rohit Ramachandran. 2016. "Population Balance Models for Pharmaceutical Processes." In *Process Simulation and Data Modeling in Solid Oral Drug Development and Manufacture*, edited by Marianthi G. Ierapetritou and Rohit Ramachandran, 43–83. New York, NY: Springer New York.

Koller, Daphne, and Nir Friedman. 2009. *Probabilistic Graphical Models: Principles and Techniques.*

MIT Press.

Maguire, Jennifer, and Daniel Peng. 2015. "How to Identify Critical Quality Attributes and Critical Process Parameters." In *Presented FDA/PQRI 2nd Conference North Bethesda, Maryland*. pqri.org. https://pqri.org/wp-content/uploads/2015/10/01-How-to-identify-CQA-CPP-CMA-Final.pdf.

Martin, Osvaldo. 2018. *Bayesian Analysis with Python: Introduction to Statistical Modeling and Probabilistic Programming Using PyMC3 and ArviZ, 2nd Edition*. Packt Publishing Ltd.

Minka, T., J. Winn, J. Guiver, Y. Zaykov, D. Fabian, and J. Bronskill. n.d. "Infer .NET 0.3, 2018. Microsoft Research Cambridge."

Qin, S. Joe. 2014. "Process Data Analytics in the Era of Big Data." *AIChE Journal. American Institute of Chemical Engineers* 60 (9): 3092–3100.

Schenkendorf, R. 2016. "Supporting the Shift towards Continuous Pharmaceutical Manufacturing by Condition Monitoring." In *2016 3rd Conference on Control and Fault-Tolerant Systems (SysTol)*, 593–98.

Steinwandter, Valentin, Daniel Borchert, and Christoph Herwig. 2019. "Data Science Tools and Applications on the Way to Pharma 4.0." *Drug Discovery Today* 24 (9): 1795–1805.

Sugiyama, Hirokazu, Yusuke Morikawa, Mai Matsuura, and Menghe Xu. 2019. "Relevance of Regulatory Constraints in Designing Pharmaceutical Manufacturing Processes: A Case Study on Waste Solvent Recovery." *Sustainable Production and Consumption* 17 (January): 136–47.

Wang, Li Ge, Ruihuan Ge, Xizhong Chen, Rongxin Zhou, and Han-Mei Chen. 2021. "Multiscale Digital Twin for Particle Breakage in Milling: From Nanoindentation to Population Balance Model." *Powder Technology*. https://doi.org/10.1016/j.powtec.2021.03.005.

Figures

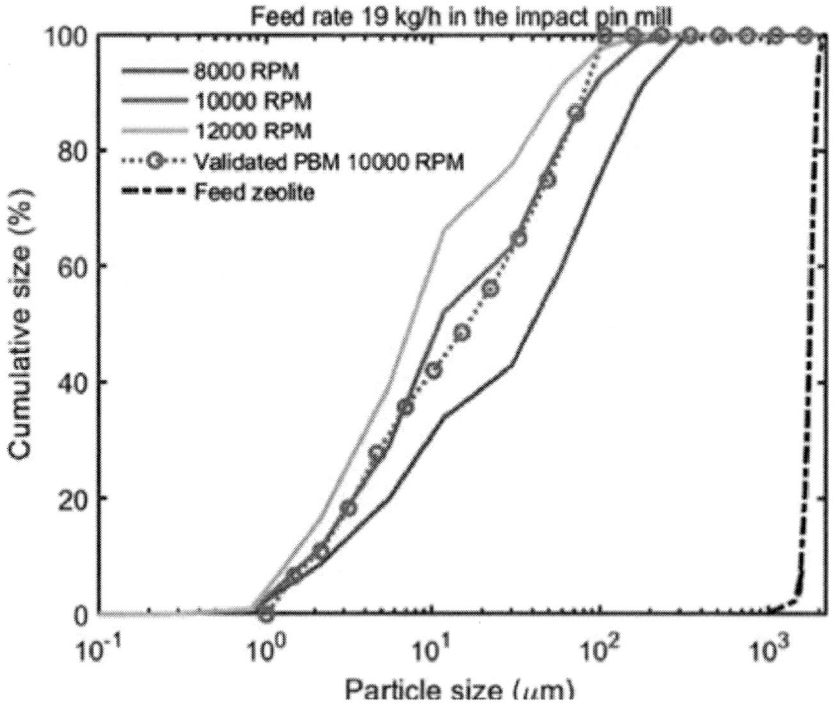

Figure 1. Particle Size Measurements from an Impact Pin Mill versus the Predictions of its Digital Twin

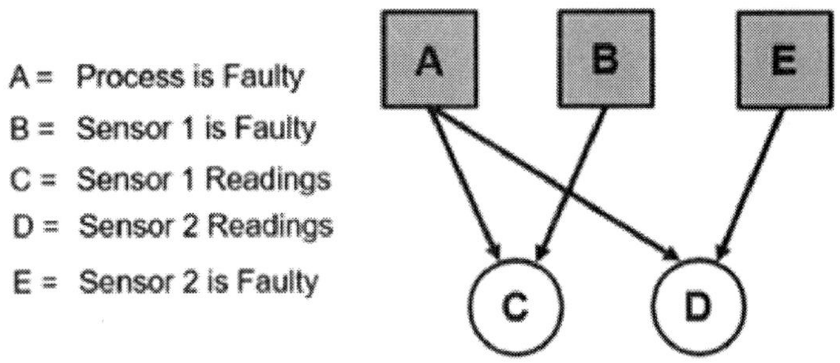

Figure 2. Directed Acyclic Graph (Bayesian Network) of a Process with Two Sensors

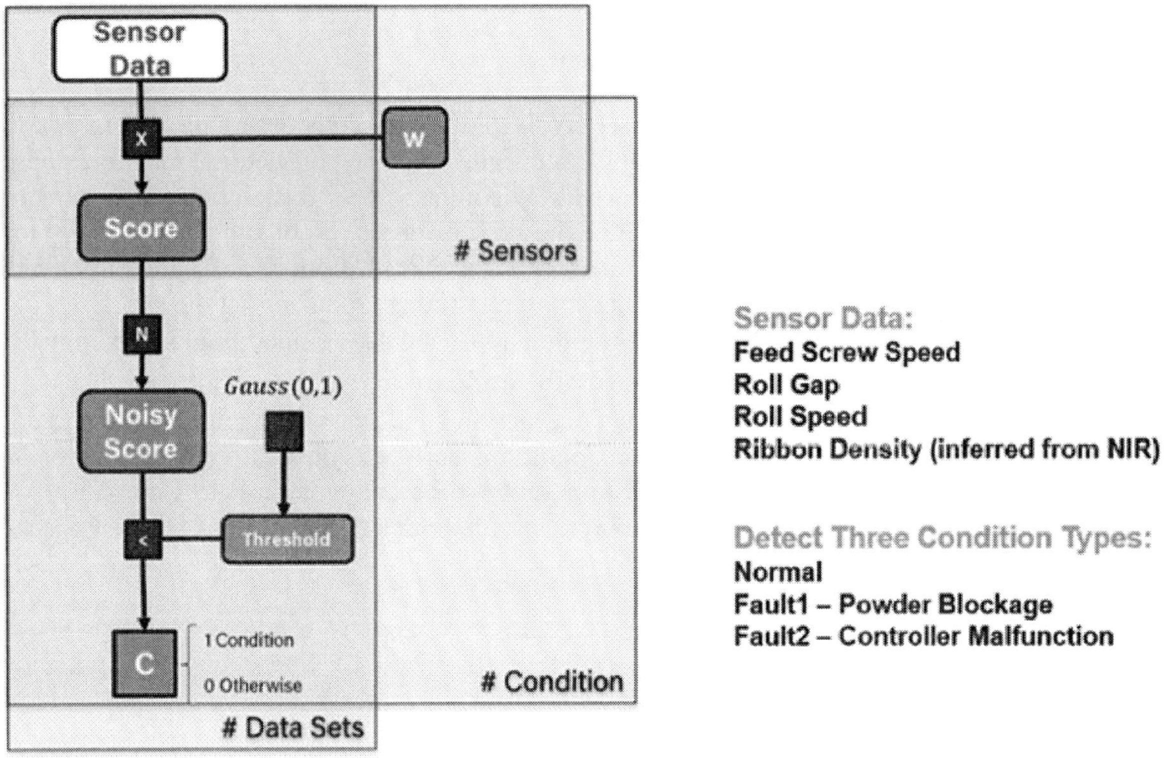

Figure 3. Probabilistic Graphical Model of a Roller Compactor Fault Detection Tool

625314 Identifying Potential One-Pot Synthesis from Computer-Planned Synthetic Routes

Abstract:

Recent years have seen rapid advancement on the development of computer-aided synthesis planning (CASP) tools. Using expert-crafted rules[1] or machine learning from chemical databases[2,3], these tools have become so powerful as to attract the attention of organic chemists, especially in the pharmaceutical industry. While good at idea generation, these CASP tools easily find numerous pathways, and prioritizing them is not a trivial task. Among the many practical considerations that a chemist would use to evaluate the synthetic routes, the possibility for one-pot synthesis development is an important aspect. Ideally in a one-pot synthesis, all reaction steps are carried out consecutively without needing to isolate intermediates. It has the benefit of minimize material loss, risk of leakage, and ease of operability.[4] To successfully develop a one-pot synthesis, it is important that the chemical species do not interfere with the subsequent reactions.

In practice, the development of one-pot synthesis involves different considerations, and in this work we explore one aspect – the similarity of reaction conditions between consecutive reaction steps. The assumption is that similar conditions will be more likely compatible with each other and not causing side reactions. While this alone does not guarantee the success of a one-pot synthesis, we aim to use this as a prioritization method at the synthesis planning stage. We explore two different approaches to identify one-pot synthesis from computer-planned synthetic routes, focusing on the similarity of reaction conditions – implicit and explicit. The implicit approach uses embedding of reaction conditions from machine learning models[5] to quantify similarity of reaction conditions. The explicit approach uses models to predict a number of sets of reaction conditions for each reaction and compare them directly. The average similarity of reaction conditions between consecutive reaction steps is used as a means of quantifying the suitability for one-pot synthesis development. A number of computational case studies are performed to qualitatively demonstrate the effect of these approaches.

Reference

1 S. Szymkuć, E. P. Gajewska, T. Klucznik, K. Molga, P. Dittwald, M. Startek, M. Bajczyk and B. A. Grzybowski, *Angew. Chemie Int. Ed.*, 2016, **55**, 5904–5937.

2 C. W. Coley, D. A. Thomas, J. A. M. Lummiss, J. N. Jaworski, C. P. Breen, V. Schultz, T. Hart, J. S. Fishman, L. Rogers, H. Gao, R. W. Hicklin, P. P. Plehiers, J. Byington, J. S. Piotti, W. H. Green, A. J. Hart, T. F. Jamison and K. F. Jensen, *Science (80-.).*, 2019, **365**, eaax1566.

3 M. H. S. Segler, M. Preuss and M. P. Waller, *Nature*, 2018, **555**, 604.

4 Y. Hayashi, *Chem. Sci.*, 2016, **7**, 866–880.

5 H. Gao, T. J. . Struble, C. W. . Coley, Y. Wang, W. H. Green and K. F. Jensen, *ACS Cent. Sci.*, 2018, **4**, 1465–1476.

632305 Optimizing Energy Efficiency of a Twin Screw Granulation Process Using a Physics-Constrained Hybrid Model

Abstract:

Digitization of manufacturing processes has led to an increase in availability of process data which has enabled the use of data-driven models to predict the process outcomes. Data-driven models are instantaneous to simulate and can provide real-time predictions but lack any governing physics within their framework. When the process deviates from original conditions, the predictions from these models may not agree with physical boundaries. In such cases, the use of first principle-based models to predict process outcomes have proven to be effective but computationally inefficient. Data-driven models unlike their first-principle counterparts do not have the ability to scale across equipment geometries. Thus, there remains a need to develop computationally efficient models with physical understanding about the processes. Training data govern the range of the prediction of the data-driven model, thus there is also a need for these models to adapt to new data, outside of the original provided domain.

In continuous oral dosage pharmaceutical manufacturing, twin screw wet granulation (TSG) is one process where such a model would be useful. Hindiyeh et al.[1] observed that blending and granulation consume about 40% of the total energy usage in downstream oral dosage pharmaceutical manufacturing. Wet granulation is an energy intensive process since a large amount of shear needs to be applied to the granulating powder blend to obtain good liquid distribution and final granules properties. During lab-scale research, a usual run time of experiments ranges from 10-15 minutes where as manufacturing lines are run longer usually from 8-16 hours. This is when monitoring of energy consumption becomes relevant. Vercruysse et al.[2] in their one-hour long experiments observed that torque required by the system gradually kept increasing till the end of their run. Several studies have optimized the TSG process for obtaining desired granule critical quality attributes (CQAs) [3], [4], but no study has addressed the potential savings in manufacturing costs by optimizing the energy efficiency of the TSG system.

In this work, twin screw granulation experiments were performed and real-time data for process parameters, equipment torque and granule quality attributes were collected using an integrated data infrastructure. This data was then transferred to cloud data lake and was used to train a long-short term memory (LSTM) network to link the real-time process parameters to torque and granule CQAs. To maximize the energy efficiency of the TSG process, the torque applied to the system by the screws needed to be minimized. A physics-constrained supervised autoencoder [5] was used to determine the domain of the process parameters in which the TSG process could be run while still maintaining the granule CQAs. The autoencoder model also predicted the granule growth regime in real-time. After a predetermined time interval, the optimization algorithm was run to determine the new process parameters for the TSG process, which in turn lead to a decrease in the torque levels of the system. This algorithm led to an increase in the energy efficiency of the system by about ~20%, leading to large potential savings in an actual manufacturing scenario.

References

[1] M. Hindiyeh *et al.*, "Process modification of pharmaceutical tablet manufacturing operations: An eco-efficiency approach," *Processes*, vol. 6, no. 2, 2018.

[2] J. Vercruysse *et al.*, "Use of a continuous twin screw granulation and drying system during formulation development and process optimization," *Eur. J. Pharm. Biopharm.*, vol. 89, pp. 239–247, 2015.

[3] H. Liu *et al.*, "Optimization of critical quality attributes in continuous twin-screw wet granulation via design space validated with pilot scale experimental data," *Int. J. Pharm.*, vol. 525, no. 1, pp. 249–263, 2017.

[4] H. Liu *et al.*, "Design space determination and process optimization in at-scale continuous twin screw wet granulation," *Comput. Chem. Eng.*, vol. 125, pp. 271–286, 2019.

[5] C. Sampat *et al., "A physics constrained supervised autoencoder model to predict critical quality attributes of a wet twin screw granulation process"* (Under preparation).

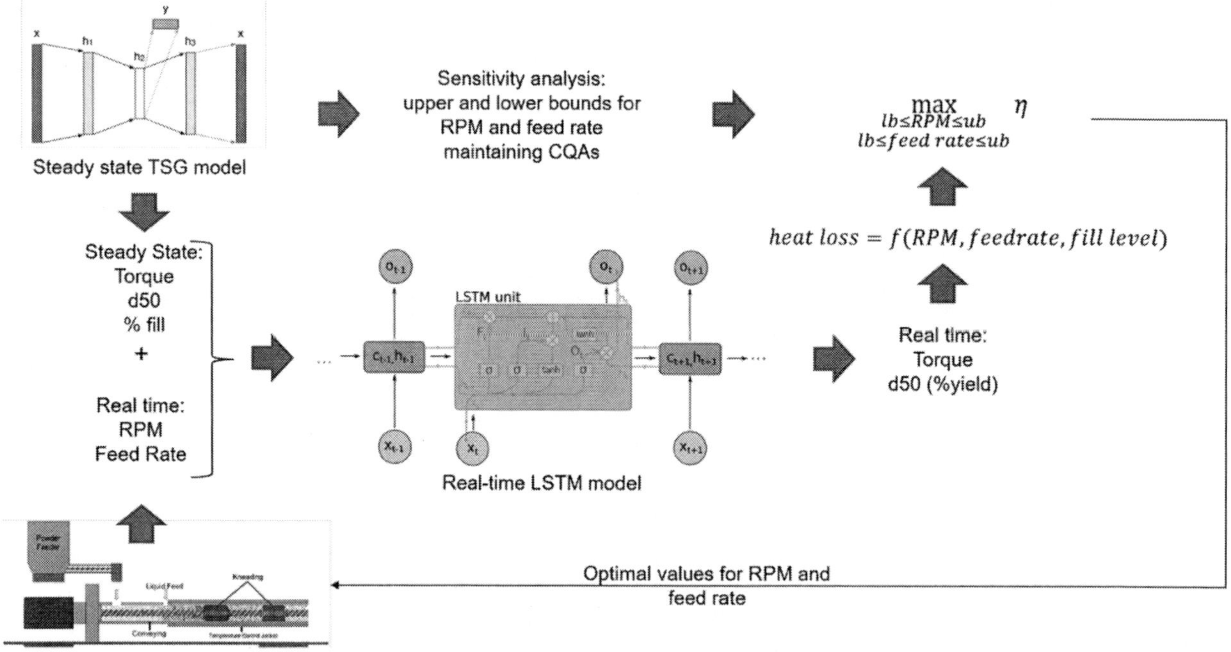

631903 Optimization and Control of Modular Chemical Systems for Continuous Manufacturing

Abstract:

Pharmaceuticals have traditionally been manufactured by using batch processing. Interest in recent years in both academia and industry has shifted towards continuous processing, with the aim to reduce drug costs, production times, waste, and product quality variations, while facilitating faster demand change responses [1,2,3]. Other recent advances in advanced manufacturing involve modular chemical systems, which employ process intensification and the potential for increased flexibility and productivity (e.g., see [4,5] and citations therein). This interest is particularly prominent in the chemicals, pharmaceuticals, biotechnology, and energy industries (e.g., [3,6,7]). This talk focuses on formulation, methodological, and computational advances in the optimization and control of modular chemical systems, with a focus on automation, to provide solutions for continuous-flow pharmaceutical manufacturing. Both first-principles-based and data-driven methodologies are discussed.

Linear MPC (LMPC) methodologies are a promising supervisory control approach for modular systems, since they have already undergone extensive testing and implementation in the chemical industries and government regulatory agencies are more likely to approve an industrial drug production facility that uses such control software. An additional advantage of LMPC formulations such as quadratic dynamic matrix control (QDMC) is the usage of input-output (IO) predictive models. IO models enable low computational cost when solving online optimizations, even for systems otherwise described by first-principles models that have very high state dimension which arise, for example, when partial differential equations are discretized (i.e., such as associated with tubular reactors in a modular system). The first part of this talk will review some methodologies for the construction of linear IO model that result in improved closed-loop performance [8], including some results that go beyond our past publication. Closed-loop simulation results for the control of a plant for the upstream synthesis of an important pharmaceutical using QDMC are presented for various scenarios of disturbances acting on the system. This talk also addresses the control of dynamical operating regions of the modular plant by employing dynamic optimization (offline) and QDMC (online) [9]. This hybrid approach results in closed-loop responses exhibiting good robustness under the existence of model parametric uncertainty for a realistic computational case study [9].

An alternative to first-principles and linear IO models are data-driven nonlinear models such as dynamic artificial neural networks (DANNs). DANNs can be used in a nonlinear model predictive control (NMPC) algorithm that is runnable in real time, which was explored in the 1980s but has received increased attention in recent years. An alternative DANN-based approach is to design approximate MPC strategies in which the control law is learned by data. Such strategies have been shown to give good closed-loop performance in simulations, and guarantees of their theoretical properties have been recently derived [10,11]. This talk discusses another alternative DANN-based approach, which employs the mathematical framework of Lyapunov theory and matrix inequalities to design dynamic state and output feedback controllers and observes for DANN models [12,13]. The methodology is demonstrated on two case studies, for a continuously stirred tank reactor control and pH control, where good closed-loop performance is observed. Given that DANNs have been used in the control of nonlinear dynamical systems in industrial practice for decades, it is conceivable that such approaches could someday become sufficiently accepted that they could be applied in the control of modular pharmaceutical manufacturing.

References

1. S. Mascia, P. L. Heider, H. Zhang, R. Lakerveld, B. Benyahia, P. I. Barton, R. D. Braatz, C. L. Cooney, J. M. B. Evans, T. F. Jamison, K. F. Jensen, A. S. Myerson, and B. L. Trout, "End-to-end continuous manufacturing of pharmaceuticals: Integrated synthesis, purification, and final dosage formation," Angewandte Chemie International Edition, vol. 52, no. 47, pp. 12359–12363, 2013.

2. I. R. Baxendale, R. D. Braatz, B. K. Hodnett, K. F. Jensen, M. D. Johnson, P. Sharratt, J.-P. Sherlock, and A. J. Florence, "Achieving continuous manufacturing: Technologies and approaches for synthesis, workup and isolation of drug substance," Journal of Pharmaceutical Sciences, vol. 104, no. 3, pp. 781–791, 2015.

3. A. S. Myerson, M. Krumme, M. Nasr, H. Thomas, and R. D. Braatz, "Control systems engineering in continuous pharmaceutical manufacturing," Journal of Pharmaceutical Sciences, vol. 104, no. 3, pp. 832–839, 2015.

4. A.-C. Bédard, A. Adamo, K. C. Aroh, M. G. Russell, A. A. Bedermann, J. Torosian, B. Yue, K. F. Jensen, and T. F. Jamison. "Reconfigurable system for automated optimization of diverse chemical reactions," Science 361(6408):1220–1225, 2018.

5. J. A. Paulson, E. Harinath, L. C. Foguth, and R. D. Braatz. "Control and systems theory for advanced manufacturing." In Emerging Applications of Control and System Theory, edited by Roberto Tempo, Stephen Yurkovich, and Pradeep Misra, Lecture Notes in Control and Information Sciences, Springer Verlag, Chapter 5, 63–80, 2018.

6. R. Lakerveld, P. L. Heider, K. D. Jensen, R. D. Braatz, K. F. Jensen, A. S. Myerson, and B. L. Trout. "End-to-end continuous manufacturing: Integration of unit operations." In Continuous Manufacturing of Pharmaceuticals, edited by P. Kleinebudde, J. Khinnast, and J. Rantanen, Wiley, New York, Chapter 13, pages 447–483, 2017.

7. M. S. Hong, K. A. Severson, M. Jiang, A. E. Lu, J. C. Love, and R. D. Braatz. "Challenges and opportunities in biopharmaceutical manufacturing control," Computers & Chemical Engineering, 110:106–114, 2018.

8. A. Nikolakopoulou, M. von Andrian, and R. D. Braatz. "Plantwide control of a compact modular reconfigurable system for continuous-flow pharmaceutical manufacturing," in Proc. American Control Conference, pp. 2158–2163, 2019.

9. A. Nikolakopoulou, M. von Andrian, and R. D. Braatz. "Fast model predictive control of startup of a compact modular reconfigurable system for continuous-flow pharmaceutical manufacturing," in Proc. American Control Conference, pp. 2778–2783, 2020.

10. A. D. Bonzanini, J. A. Paulson, D. B. Graves, and A. Mesbah, "Toward safe dose delivery in plasma medicine using projected neural network-based fast approximate NMPC," in Proc. IFAC World Congress, pp. 5353–5359, 2020.

11. H. H. Nguyen, T. Zieger, S. C.Wells, A. Nikolakopoulou, R. D. Braatz, and R. Findeisen, "Stability certificates for neural network learning-based controllers using robust control theory," in Proc. American Control Conference, in press, 2021.

12. A. Nikolakopoulou, M. S. Hong, and R. D. Braatz, "Feedback control of dynamic artificial neural

networks using linear matrix inequalities," in Proc. IEEE Conference on Decision and Control, pp. 2210–2215, 2020.

13. A. Nikolakopoulou, M. S. Hong, and R. D. Braatz, "Output feedback control and estimation of dynamic neural networks using linear matrix inequalities," in Proc. American Control Conference, in press, 2021.

622636 Developing a Workflow for Continuous Centrifugal Extraction

Abstract:

Liquid-liquid extraction is ubiquitous throughout API process development as a method to purify reaction mixtures. On paper, extraction is a straightforward, effective unit operation in which two immiscible liquids are mixed to extract specific components from one liquid phase to the other. In practice, extractions can pose problems as they are scaled up from development work to manufacturing scales. Traditionally, extraction in the pharmaceutical industry is performed in batch mode in stirred tanks, which can lead to long settling times and inefficient use of solvent. Furthermore, the presence of surfactants or the over-agitation of the two liquids may lead to emulsification, which significantly complicates the gravity-driven separation of the two liquids. Prolonged settling times increase the cycle time of the process, reducing overall throughput and may risk degrading reactive intermediates as the liquids remain in contact while settling.

Continuous extraction offers attractive, practical solutions to conventional batch extraction. First, continuous flow devices have smaller footprints compared to batch devices. Additionally, there are many types of flow contacting devices, such as mixer-settlers, packed or agitated columns, and membranes, among others, allowing flexibility for the contacting device to match the properties and needs of the two liquids in the process. One such continuous flow contacting device is a centrifugal extractor. These devices provide excellent mixing for the two liquids as they flow into the device while efficiently separating the two liquids via centrifugal force as they flow out of the device. Centrifugal extractors offer mixing and separation in the same unit to minimize the footprint, can separate liquids with small density differences, can tolerate liquids that tend to emulsify, and provide short contacting times for fast extractions or unstable intermediates. These advantages of centrifugal extractors make them attractive as options for continuous flow in the pharmaceutical industry.

This presentation describes the development work of a model system for continuous centrifugal extraction. We will describe the key features of the process, such as partition coefficient, number of equilibrium stages, and other relevant thermodynamic data. Additionally, we will highlight some of the key issues and parameters to address with scale-up and design of centrifugal extraction processes. Finally, we will detail AbbVie's general workflow to designing and developing continuous extractions with the goal of implementing robust continuous extractions at scale.

Eric Moschetta and Benjamin Rizkin are employees of AbbVie and may own AbbVie stock. AbbVie sponsored and funded the study; contributed to the design; participated in the collection, analysis, and interpretation of data, and in writing, reviewing, and approval of the final publication.

629625 A Scalable Continuous Reaction and Isolation Process for the Production of Sulfonyl Chloride Pharmaceutical Intermediates

Abstract:

Chlorosulfonic acid is essential for many pharmaceutical processes due to its ability to selectively chlorosulfonate aryl rings via an S_E2 mechanism (*1*). The chlorosulfonates are precursors to a wide range of biologically active molecules including sulfonylureas (*2*) and sulfonamides (*3*). The hazardous and corrosive nature of chlorosulfonic acid is challenging to address particularly during scale up for an industrial manufacturing process (*1*). The approach taken to address these challenges was to perform this reaction in two consecutive continuous stirred tank reactors followed by a continuous isolation. The implementation of continuous manufacturing can be advantageous to improve the safety, speed, product quality, and environmental footprint of challenging reactions (*4*). The safety advantages of transitioning this reaction to continuous manufacturing include a reduced amount of heated chlorosulfonic acid, the gradual release of toxic gases, and limited operator exposure. The transition to a continuous isolation step can greatly improve space-time yield as the slow addition of the final product mixture to the isolation vessel was determined to be an essential parameter for maintaining the critical quality attributes.

Following the development of preliminary reaction and quenching conditions, multiple 100-g scale flow experiments were performed using two continuously stirred round bottom flasks, each with a one hour mean residence time. Peristaltic pumps were used with a combination of C-Flex and PTFE tubing to transfer the reaction mixture between each vessel. The initial smaller scale flow runs revealed a need for process monitoring of CSTR levels so that flow rates could be adjusted as needed. Mettler Toledo scales were connected to a Python terminal via a serial port connection and a custom script was implemented so that changes in relative levels of the starting material and each round bottom could be detected and recorded in real time. The isolation step was performed with periodic rapid pumping from the quenching vessel into the filtration step to prevent excessive solid product accumulation.

A 500-g continuous manufacturing run was next performed with a 20-L jacketed reactor used for the precipitation step. A high-purity sulfonyl chloride product was obtained in moderate yield as determined by HPLC assay, ^{1}H NMR, and LCMS. Process monitoring data from the Python script revealed that the increased flow rate of the larger scale manufacturing run led to problems with C-Flex tubing compatibility and to subsequent variable flow rates and skewed residence times. Peristaltic pump heads for PTFE tubing were therefore purchased and implemented for all future flow experiments. Further batch-scale chemistry development elucidated that a higher number of equivalents of chlorosulfonic acid was needed to react with the sulfonic acid intermediate and that the initial work-up should be done at sub-ambient conditions. In order to improve green chemistry metrics, experiments were conducted minimizing the amount of water used during the isolation step. Lastly, the recommended process improvements were implemented in 100-g scale and 500-g scale continuous manufacturing runs.

Literature Cited

1. **Cremlyn, R.J.,** "Chlorosulfonic Acid: A Versatile Reagent," Royal Society of Chemistry,

Cambridge, UK (2002).
2. **Sen S., Ruchika, Kumar D., Easwari T.S., and S. Gohri,** "Therapeutic Aspects of Sulfonylureas: A Brief Review," *Journal of Chemical and Pharmaceutical Research,* **8** (12), pp. 121-130 (Dec. 2016).
3. **Kolaczek, A., Fusiarz, I., Lawecka, J., and D. Branowska,** "Biological Activity and Synthesis of Sulfonamide Derivatives: A Brief Review," *CHEMIK,* **68** (7), pp. 620-628 (2014).
4. **Baumann, M., Moody, T.S., Smyth, M., and S. Wharry,** "A Perspective on Continuous Flow Chemistry in the Pharmaceutical Industry," *Process Res. Dev.,* **24** (10), pp. 1802–1813 (2020).

626756 Target Polymorphic Form Development Via Continuous Combined Cooling and Antisolvent Crystallization Using Oscillatory Baffled Crystallizer

Abstract:

In recent years, continuous manufacturing has received much interest across the pharmaceutical industry, academia and regulatory bodies due to its potential to lower the cost of manufacturing while maintaining consistent quality[1-3]. The regulatory bodies, such as the U.S. Food and Drug Administration (FDA), have been encouraging pharmaceutical companies to adopt continuous manufacturing to ensure uninterrupted supply of high-quality drugs to patients[4]. One of the important links in continuous manufacturing is crystallization unit operation. Most often crystallization governs the critical quality attributes of active pharmaceutical ingredient (API) such as purity, crystal size distribution (CSD), yield, polymorphic form, morphology, etc. In continuous crystallization, the controlled state of operation helps in achieving consistent quality drug substance. Continuous cooling only crystallization on model compounds has been extensively studied[5-6], but a very few papers have been published on continuous combined cooling and antisolvent crystallization (CCAC). It is common in pharmaceutical industry to use antisolvent to improve yield, but the solvent composition can impact the polymorphic outcome of the drug substance[7]. Therefore, designing a continuous CCAC process to solve the challenges of API crystallization in a highly regulated environment is a complex engineering problem. In this work, continuous CCAC of a commercial drug substance, Atorvastatin calcium, via oscillatory baffled crystallizer will be presented.

Atorvastatin calcium (ASC) is a statin medication that is used as a lipid lowering agent and to prevent cardiovascular diseases. It works by inhibiting HMG-CoA reductase, an enzyme found in liver tissue that plays a key role in production of cholesterol in the body. With cardiovascular diseases being the number one cause of death in the world[8], the demand for the statin drugs is quite high, making it a high-volume drug substance and hence a suitable candidate for continuous manufacturing.

ASC is currently manufactured via batch CCAC, which is time intensive and gives wide CSD product at the end of the batch. To overcome this and other disadvantages of batch crystallization, continuous CCAC of ASC was studied in an Alconbury Weston Ltd DN-15 oscillatory baffled crystallizer (OBC) with the aim to improve productivity and CSD of the desired polymorph. An OBC has the advantage of high heat transfer rates and improved mixing that significantly brings down the crystallization time. It also has the advantage of spatial temperature distribution and multiple antisolvent addition points to control supersaturation and hence the crystallization process.

The desired polymorph of ASC form I is crystallized from the mixture of isopropyl alcohol (IPA) and water. First the solubility of ASC form I was found at different temperatures and solvent compositions. The solubility increased as the temperature increased but the solvent composition had a synergistic effect on the solubility. As the desired form is a trihydrate, the saturated API solution was prepared in a high IPA:water ratio and the final solvent composition had low IPA:water ratio.

Three different modes of batch CCAC process were studied to understand the effects of temperature and solvent compositions on the polymorphic outcome. The first mode was direct antisolvent addition in which the antisolvent added to the seeded saturated API solution. In the second mode, reverse antisolvent addition was carried out in which saturated API solution was added to seeded antisolvent.

In the third mode, distributed antisolvent addition was tested in which the antisolvent was split into two parts (AS-1 and AS-2). The saturated API solution was added in the seeded AS-1 and then the remaining AS-2 was added to system. The first mode did not yield the desired polymorph, but the second and the third modes gave the desired form. In the first mode, the IPA rich solution did not favor the desired form nucleation, whereas in the other two modes, the IPA concentration was much lower which favored the nucleation of the desired form.

In case of continuous crystallization, there is no direct or reverse antisolvent crystallization as the saturated API and antisolvent streams mix in the OBC tube at a point irrespective of which is added first. However, the ratio of IPA:water at the point of mixing of the two streams was found to have an impact on the polymorphic outcome and the operation in the OBC. The presence of seeds in either saturated API solution or antisolvent also impacted the polymorphic outcome. The splitting of antisolvent into two addition ports in the OBC was found to give the desired form. Through enhanced micromixing and increased heat transfer characteristics of the OBC, the continuous process was found to be 70-fold more productive compared to the batch process. The crystals obtained from continuous OBC operation had narrower CSD compared to that from a batch CCAC process.

Acknowledgement

Dr. Reddy's Laboratories Ltd. (DRL) is acknowledged for financial support provided through Purdue – DRL fellowship program.

Reference:

1. McKenzie, P., Kiang, S., Tom, J., Rubin, A. E. & Futran, M. Can Pharmaceutical Process Development Become High Tech? AIChE J. 52, 3990–3994 (2006).
2. Plumb, K. Continuous processing in the pharmaceutical industry: Changing the mind set. Chem. Eng. Res. Des. 83, 730–738 (2005).
3. Lee, S. L., O'Connor, T. F., Yang, X., Cruz, C. N., Chatterjee, S., Madurawe, R. D., Woodcock, J. Modernizing Pharmaceutical Manufacturing: from Batch to Continuous Production. *Journal of Pharmaceutical Innovation*, *10*(3), 191–199 (2015).
4. Pharmaceutical CGMPs for the 21s Century - A risk-based approach. Food Drug Adm. 32 (2004).
5. Chew, C. M., Ristic, R. I., Dennehy, R. D., & De Yoreo, J. J. Crystallization of paracetamol under oscillatory flow mixing conditions. Cryst. Growth Des., 4(5), 1045-1052 (2004).
6. Ristic, R. I. Oscillatory mixing for crystallization of high crystal perfection pharmaceuticals. Chem. Eng. Res. Dev., 85(7), 937-944 (2007).
7. Lindenberg, C., Krattli, M., Cornel, J., Mazzotti, M. Design and Optimization of a Combined Cooling/Antisolvent Crystallization Process. Cryst. Growth Des. 9(2), 1124–1136 (2009).
8. Global Burden of Disease Collaborative Network. Global Burden of Disease Study 2016 Results. Seattle, United States: Institute for Health Metrics and Evaluation (IHME) (2017).

Publication no.: IPDOIPM-00657

627822 Development of a Novel Continuous Spatially Distributed Diafiltration Unit Operation

Abstract:

Flow chemical and continuous processes have seen significant uptake in R&D and small intensified manufacturing for complex chemical and pharmaceutical products. However, continuous work-up and purification operations for continuous manufacturing requires further development to approach the significant level of purification and stream adjustment in multistep routes that batch synthesis traditionally offers in the production of these classes of chemical products.

The objective of this study is to develop an operation that can conduct separations based on diafiltration using semipermeable nanofiltration or ultrafiltration membranes in a fully continuous manner in a single stage configuration. To this end a Continuous Spatially Distributed Diafiltration (CS2D) operation was developed, to conduct straight through diafiltration in a manner that would yield equivalent or better purification efficiency, yield and productivity to batch diafiltration. To achieve this goal diafiltration solvent is introduced spatially in a highly uniform manner with the flow guided by a range of 3D-printed mixers designed using CFD informed design to increase localized mixing of retentate and diavolumes flows or displacement effects within the channel. Static mixers were 3D-printed using titanium and polyether ether ketone (PEEK) providing high pressure and chemical compatibility (Figure 1), suitable for many flow chemical separations.

Figure 1: (a) mixer filled membrane flatform (1-bottom plate; 2- middle plate to hold membrane with double O-ring; 3-mixer; 4-top plate) and (b) selection of printed mixers

Ibuprofen was selected as a model active pharmaceutical ingredient (API) with methanol acting as a solvent and ethanol used as a model impurity where an organic solvent nanofiltration membrane was used to selectively retain ibuprofen. Experiments were processed at a 55 bar transmembrane pressure at which 87% and 15% rejection coefficients of ibuprofen and ethanol were determined respectively, in screening experiments. Using 4 diavolumes in constant volume diafiltration in batch and CS2D runs, purities of 55% and 76% were obtained at equivalent yields of approximately 61% for the Batch and CS2D separation respectively. Demonstrating that continuous spatially distributed diafiltration (CS2D) can match or exceed results conducted in equivalent batch configurations. As such CS2D may present an attractive intermediate purification and solvent swap operation for telescoped flow chemical and continuous processing applications, in addition to a highly compatible platform for use in automated flow-based experimentation.

Acknowledgements

This publication is supported SSPC, the Science Foundation Ireland Research Centre for Pharmaceuticals (12/RC/2275_P2) and I-form the Science Foundation Ireland Centre For Advanced manufacturing (16/RC/3872) and TIDA, SFI/EI Technology Innovation Development award (16/TIDA/4003) co-funded under the European Regional Development Fund.

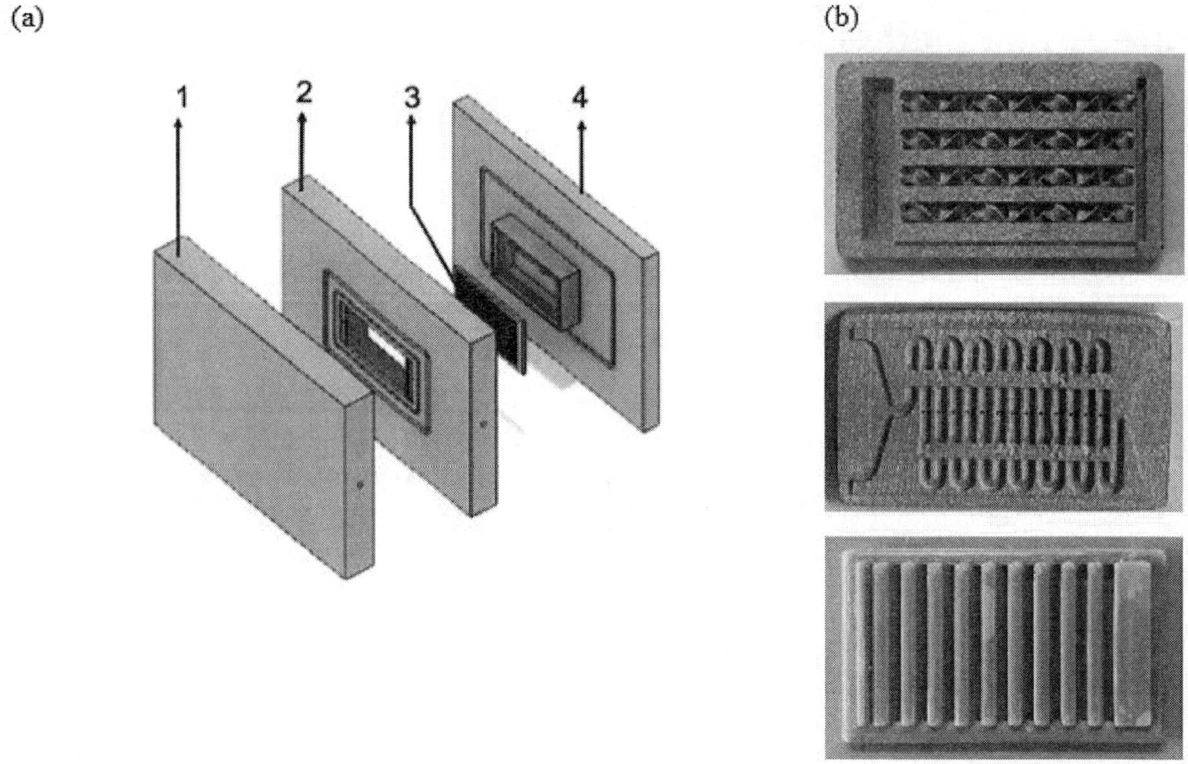
(a)
(b)
1
2
3
4

624587 Predicting Screw Feeder Flow Rates from Powder Properties and Operating Conditions

Abstract:

A well-known bottleneck in drug product development is the dependence on experimentally fit semi-empirical flowsheet models to design continuous manufacturing processes. The creation of predictive flowsheet models that account for powder properties and operating conditions effects on powder flow loosens this bottleneck by enabling engineers to begin design upon characterization and inform experimentation, ultimately speeding up a product's time-to-market. Such processes begin with identifying several screw feeders which are responsible for controlling the mass flow rates of active pharmaceutical ingredients and excipients across the entire process. Prior work has shown screw feeder mass flow rates are related to both the feeder configuration [1]–[3] and a powder's characterized properties [4]–[8], but do not provide a framework for predictively simulating a screw feeder's dynamics. Our paper addresses this process development gap by proposing a predictive, high-fidelity flowsheet model that can provide insight into how powder characteristics, feeder configuration, and operating state influence a screw feeder's dynamic behavior.

Our predictive feeder model is built from three components: a simple volumetric displacement screw feeder flowsheet model, a statistical autoregressive moving average model (ARMA), and a predictive projection to latent spaces (PLS) deterministic model. First, the hybrid deterministic displacement and stochastic ARMA model, as proposed by Johnson et al. [9], were fit to experimental runs of five powders with varying property values and feeding behaviors using both Coperion K-tron KT20 and MT12 feeders with twin concave screws operated at many different volumetric and gravimetric setpoints. This yielded sets of deterministic and stochastic deviation model parameters for each powder-feeder-setpoint operating configuration. Next, a PLS model was built that maps a database of powder properties [10], an abstraction of feeder geometries, and operating state to their experimentally estimated deterministic and stochastic parameters, extending the work of Bostijn et al. [8]. Constraints were integrated into the PLS model as needed to ensure deterministic parameters are physically possible and stochastic ARMA parameters meet theoretical requirements for stability and invertibility. The PLS model scores and loadings are presented, providing insight into how different powder properties influence the minute-by-minute and second-by-second dynamic feeding behavior. Last, the predictive capabilities of our PLS model are demonstrated and evaluated by simulating the feeding behavior of out of sample powders and conditions.

[1] H. Berthiaux, K. Marikh, and C. Gatumel, "Continuous mixing of powder mixtures with pharmaceutical process constraints," *Chemical Engineering and Processing: Process Intensification*, vol. 47, pp. 2315–2322, 2008.

[2] J. W. Fernandez, P. W. Cleary, and W. McBride, "Effect of screw design on hopper drawdown of spherical particles in a horizontal screw feeder," *Chemical Engineering Science*, vol. 66, pp. 5585–5601, 2011.

[3] W. E. Engisch and F. J. Muzzio, "Method for characterization of loss-in-weight feeder equipment," *Powder Technology*, vol. 228, pp. 395–403, 2012.

[4] L. X. Liu, I. Marziano, A. C. Bentham, J. D. Litster, E. T. White, and T. Howes, "Effect of particle properties on the flowability of ibuprofen powders," *International journal of pharmaceutics*, vol. 362,

pp. 109–117, 2008.

[5] M. Capece, K. R. Silva, D. Sunkara, J. Strong, and P. Gao, "On the relationship of inter-particle cohesiveness and bulk powder behavior: Flowability of pharmaceutical powders," *International Journal of Pharmaceutics*, vol. 511, pp. 178–189, 2016.

[6] Y. Wang, T. Li, F. J. Muzzio, and B. J. Glasser, "Predicting feeder performance based on material flow properties," *Powder Technology*, vol. 308, pp. 135–148, 2017.

[7] U. V. Shah, V. Karde, C. Ghoroi, and J. Y. Y. Heng, "Influence of particle properties on powder bulk behaviour and processability," *International journal of pharmaceutics*, vol. 518, pp. 138–154, 2017.

[8] N. Bostijn, J. Dhondt, A. Ryckaert, E. Szabó, W. Dhondt, B. Van Snick, V. Vanhoorne, Ch. Vervaet, and Th. De Beer, "A multivariate approach to predict the volumetric and gravimetric feeding behavior of a low feed rate feeder based on raw material properties," *International Journal of Pharmaceutics*, vol. 557, pp. 342–353, 2019.

[9] B. J. Johnson, S. García-Muñoz, M. Sen, J. Hanson, D. Slade, and N. V. Sahinidis, "(366d) stochastic analysis and modeling of pharmaceutical screw feeder mass flow rates," in *Proceedings of the 2020 Virtual AIChE Annual Meeting*, 2020.

[10] B. Van Snick, J. Dhondt, K. Pandelaere, J. Bertels, R. Mertens, D. Klingeleers, G. Di Pretoro, J. P. Remon, Ch. Vervaet, Th. De Beer, V. Vanhoorne, "A multivariate raw material property database to facilitate drug product development and enable in-silico design of pharmaceutical dry powder processes," *International journal of pharmaceutics*, vol. 549, pp. 415–435, 2018.

631153 Model-Based Analysis of Breakage in Fluid Bed Drying of Continuously Produced Pharmaceutical Wet Granules

Abstract:

In continuous pharmaceutical wet granulation, subsequent drying is a necessary step for evaporating the granulation liquid from the granules and rendering these processable downstream and in tableting. In the ConsiGma™-25 wet granulation this is achieved through semicontinuous fluid bed drying. The desired effect of drying to the ideal moisture content is accompanied by breakage of granules during this process (De Leersnyder et al., 2018). The generated fines have a negative impact on the processability downstream, such as possibly reduced content uniformity and flowability, as well as to the drying stage itself where the fine material gets stuck on the filter that entrains particulate material in the outgoing air. The latter could affect the duration limit of continuous operation as the pressure across the filters increases due to the fouling.

Not only does this breakage cause lower processability, it is also complex to capture and predict. It namely occurs in different locations in the process, within each of a different nature. Wet granules exiting the granulator are sent to the dryer through gravitational loading in a vertical ConsiGma™ setup, whereas in horizontal configurations these are pneumatically conveyed through a tube. Through contact with the tube wall granules undergo size reduction, dependent on their current properties governed by composition, size and moisture content. The liquid moreover provides the granules with different elastic properties than in a dry state. The next stage is during the fluid bed drying itself, as granules are dropped into the fluidizing air from the top of the dryer and collide with the dryer cell walls or already present granules. This breakage extent is affected by dryer process settings and granule properties. Finally, often the largest size reduction of the product occurs in the discharge of the dried granules to the subsequent evaluation module. A more forceful pneumatic conveying transports the granules through another tube, termed dry transfer line, causing breakage in function of granule properties, most prominently remaining moisture content.

As this problem is too complex to forecast, mathematical models can be used to predict the extent and nature of breakage occurring due to location, material and granule properties. A change in particle size distribution (PSD) can be modeled by means of a population balance model (PBM). Breakage models can be implemented in this framework, with kernels describing (combinations of) different fragmentation mechanisms that are plausible for the system at hand. The extent of breakage and other parameters of these functions can be mathematically linked to the modeled materials' properties by using PSD data.
The quality of such developed models is however not only dependent on the supplied data, also applying the right objective function plays a key role in finding the model and mechanism that describes the system in the most robust way. PSD data contains an abundance of information, it is therefore difficult to find an objective function that can take all of it into account. Also the experimental variability of PSD results could be used in model calibration in order to emphasize the most certain experimental information in the distributions. This work therefore implements techniques from the compositional data field to calculate variances on experimental PSD outcomes in function of size class. The (inverse of) the measurement uncertainty is then taken into account in the objective function. It has been shown that this allows to include information in both number- and volume-based PSD for mechanistic purposes (Ghijs, 2020).

Recent data collections have been carried out on the ConsiGma™-25 that varied the active pharmaceutical ingredient (API) and its powder properties with a fixed combination of excipients of 30% lactose monohydrate (Pharmatose 200M, DFE Pharma, Veghel, The Netherlands), 15% MCC (Avicel® PH-101, FMC International Health and Nutrition Cork, Ireland) and 5% HPMC (Methocel™ E15, The Dow Chemical Company, Midland, U.S.A.). The applied APIs showed different degrees of hydrophilicity and solubility. For each API, granules were produced at different levels of liquid addition. Part of the study also involved producing granules of similar properties based on different APIs, in order to isolate the effect of the applied API. Drying experiments were then carried varying the dryer inlet air temperature, mass loading and drying time. The experimental outcomes were characterized by high resolution PSD measurements (dynamic image analysis, QICPIC, Sympatec Gmbh, Clausthal-Zellerfeld, Germany) of the incoming and outgoing granules of the dryer.

In combination with the breakage PBM including measurement uncertainty in model calibration and validation, this study elucidates the effect of the tested API properties, granule properties (granule size and moisture content) and dryer process settings on their breakage behavior through the ConsiGma™-25 system. The more robust calibration helps in picking out the most appropriate breakage kinetics out of the abundance of PSD data. Moreover, a predictive model is generated that can take process settings, material and granule properties into account in forecasting breakage over the dryer for the selected formulations. This constitutes a pivotal modelling step towards a formulation-generic breakage model.

References

De Leersnyder, F., Vanhoorne, V., Bekaert, H., Vercruysse, J., Ghijs, M., Bostijn, N., Verstraeten, M., Cappuyns, P., Van Assche, I., Vander Heyden, Y., Ziemons, E., Remon, J. P., Nopens, I., Vervaet, C., & De Beer, T. (2018). Breakage and drying behaviour of granules in a continuous fluid bed dryer: Influence of process parameters and wet granule transfer. *European Journal of Pharmaceutical Sciences*, *115*, 223–232. https://doi.org/10.1016/j.ejps.2018.01.037

Ghijs, M. (2020). *Modelling and experimentation on fluid bed drying of granules in the context of pharmaceutical continuous wet granulation*. Universiteit Gent. Faculteit Bio-ingenieurswetenschappen.

629342 Development of a DEM-Based Digital Twin of a Continuous Direct Compression Line

Abstract:

A Continuous Direct Compression line (CDC) is a minimal setup for the production of solid dosage forms and consists only of three unit operations: feeding of the individual components, mixing the individual material streams, and compressing the blend into tablets. The discrete element method (DEM) is a powerful tool for in-depth investigations of the individual unit operations, however, the computational effort does not allow to model a complete CDC line at once.

In order to create digital twin of the CDC line, reduced order models are needed. A popular implementation of the reduced order models is based on residence time distributions (RTDs) coming from tracer experiments [1] or simulations [2]. The digital twin is able to predict the content uniformity after each processing step and in the final tablets with real feeder fluctuation data and under constructed disturbance scenarios.

In addition to exemplary scenarios, the digital twin is able to create so-called funnel plots. The funnel plots show the maximum duration and intensity of disturbances that still allow the production of tablets with in-spec content uniformity without any control action. The digital twin is thus a valuable tool to develop the process control strategy for the CDC line.

References

[1] Lee et al., 2021. " Continuous Mixing Technology: Characterization of a vertical mixer using residence time distribution". Int J Pharm, in press. https://doi.org/10.1016/j.xphs.2021.01.035

[2] Toson et al. "Continuous Powder Mixing Technology: Validation of the DEM Model". submitted to J Pharm Sci.

629679 Enabling Collaboration and Accelerating Model Deployment Via a Cloud Based Computing and Software As a Service Model

Abstract:

It has long been recognized that many aspects of model generation and implementation in the domain of pharmaceutical process development belong in the pre-competitive space. Motivated by the high value that this type of collaboration could generate;, the Modeling Group of the Enable Technology Consortium completed a project in 2019 to partner with a provider of cloud HPC services (Rescale) incorporating a software-as-a-service model. This presentation will discuss the modalities in which the modeling platform has been leveraged to foster intra- and inter-company collaborations in the construction and deployment of mathematical models and the resulting increase in their adoption across different stages of the pharmaceutical process development cycle. The use of cloud-based computing resources combined with an easy-to-use and collaborative web-based interface coupled with automation tools like Rest-APIs promises to dramatically increase the direct access of end users to software applications which would normally be inaccessible due to their high computational requirements. Finally, we discuss the anticipated evolution of the platform to adapt to a highly dynamic landscape in cloud HPC and SaaS.

624807 Uncertainty Estimation in Drug Substance Process Development with Mechanistic Models

Abstract:

Understanding and defining the design space with statistical and mechanistic methods is the cornerstone for quality by design (QbD) in pharmaceutical process development. Using probabilistic (Bayesian) models to estimate the risk of failure within the design space has been an emerging approach for risk-based control of the design space. In this work, we implemented an adaptive mechanistic model in a full Bayesian framework for design space mapping. Compared to traditional statistical models, this approach allows the simultaneous estimation of multiple CQAs and their associated uncertainty estimations. This framework can be easily adapted to different chemical transformations through a customized interface with Reaction Lab™ from Scale-Up Systems, lowering the barrier for wider use by other scientists. This model was applied to recent portfolio projects. The data supporting the model was collected from a time-resolved DOE, which allows comparison of the mechanistic Bayesian model to a statistical Bayesian model. This approach allows a risk-based evaluation and confirmation of the existing proven acceptable ranges (PARs) to assure smooth tech transfer of the process to a new site.

626412 Silica Adsorption Vaccine Encapsulation: A Case Study of Invasion Plasmid Antigen D

Abstract:

It is estimated that 50% of vaccines produced annually are wasted because effectivity is dependent on protein structure and heat exposure disrupts the intermolecular interactions needed to maintain structure. Since 90% of vaccines require a temperature-controlled supply chain, it is necessary to create a cold chain system to minimize vaccine waste. A more sustainable technology was developed via the adsorption of invasion plasmid antigen D (IpaD), onto mesoporous silica, improving the thermal stability of this protein-based therapeutic. Multiple silicas were characterized to determine the effects of pore diameter, pore volume, and surface area on protein adsorption. The silica-IpaD complex was then heated above the IpaD denaturing temperature and *N*,*N*-dimethyldodecylamine *N*-oxide (LDAO) was used to remove IpaD from the silica. Circular dichroism (CD) confirmed that the adsorbed IpaD after the heat treatment maintained its native secondary structure rich in α-helix content. In contrast, the unprotected IpaD after heat treatment lost its secondary structure. Isotherms modeled using Langmuir, Freundlich and Temkin models demonstrated that the adsorption of IpaD onto silicas is best fit by the Langmuir model. If pores are less than 15 nm, adsorption is negligible. If the pores are between 15-25 nm, then monolayer coverage is achieved and IpaD is protected from thermal denaturing. If pores are larger than 25 nm the adsorption is multilayer coverage and it is easier to remove the protein from the silica due to a less developed hydrogen bond network. This case study provides strong evidence that IpaD is thermally stabilized via adsorption on mesoporous silica with the proper pore size range.

631878 Morphology prediction for crystals: hybrid force field and COSMO-based solvation free energy model

Abstract:

Crystal morphology plays a significant role in many industrial applications, with different crystal habits possessing different physical and chemical properties. However, due to the large number of crystals, coupled with numerous growth conditions including solvent, additive, temperature, supersaturation, and type of crystallization, it is a challenging task for the industry to obtain desired crystals with particular product functionality and morphology. Therefore, an academic design software aid called ADDICT (Advanced Design and Development of Industrial Crystallization Technology) was developed by our group at UCSB to predict crystal morphology [1].

Recently, we have upgraded ADDICT's software framework [2] and completely rewritten the code based on Matlab using object-oriented programming in ADDICT3 (the third version of ADDICT) [3]. Based on the new software framework, ADDICT3 can interpret crystallography of varying types of crystals (including organic crystals, organic salts, solvates, and metal organic crystals) with varying Z'. Therefore, theoretically, we can use ADDICT3 to try to predict the morphology of a wide variety of organic crystals.

Force fields are necessary for calculating interaction energies to predict the crystal morphology using mechanistic models. We previously used the Generalized Amber Force Field [4] (GAFF 1.8) to calculate solid-state energy for organic crystals. However, the GAFF force field cannot handle crystals containing metal atoms, such as metal organic crystals. The Universal force field (UFF) [5] is an all-atom potential that has parameters for every element from hydrogen to lawrencium. But the results of UFF are sometimes not so accurate [6].

In this study, based on the advantages of GAFF and UFF, we proposed the Hybrid Force Field (HFF) to predict the morphology for metal and metal-free organic crystals. Three successful cases on the morphology prediction of crystals grown from vapor have been studied. They are (1) olanzapine, (2) metal-free phthalocyanine (H_2Pc), and (3) β-Polymorphs of copper phthalocyanine (CuPc), respectively. The results demonstrate the reliability of the HFF proposed in this paper. In addition, UFF and HFF have been incorporated into the ADDICT software, which can be used to predict the morphology of organic crystals containing any relevant type of atoms.

References

[1] Li, J., Tilbury, C. J., Kim, S. H., & Doherty, M. F. (2016). Prog. Mater. Sci., 82, 1.

[2] Zhao, Y., Tilbury, C. J., Landis, S., Sun, Y., Li, J., Zhu, P., & Doherty, M. F. (2020). Cryst. Growth Des., 20, 2885.

[3] Landis, S., Zhao, Y., & Doherty, M. F. (2020). Comput. Chem. Eng., 133, 106637.

[4] Wang, J., Wolf, R. M., Caldwell, J. W., Kollman, P. A., & Case, D. A. (2004) J. Comp. Chem., 25, 1157.

[5] Rappé, A. K., Casewit, C. J., Colwell, K., Goddard III, W. A., & Skiff, W. M. (1992). J. Am. Chem. Soc., 114, 10024.

[6] Martin, M. G. (2006). Fluid Phase Equilibria, 248, 50.

629714 Analysis of a Natural Compound Therapeutic on Engineered Ovarian Cancer Spheroids

Abstract:

Ovarian cancer the most lethal gynecological cancer with fatality to higher than 70% of patients in 5 years after diagnosis[1]. Tumor cells shed from the primary tumor as single or multicellular and form spheroids within the peritoneal cavity[2]. Spheroid formation within patient ascites increase cancer cell resistance against anoikis[3]and anti-cancer drugs[4,5]. The clinical outcome of ovarian cancer has led to the development of three-dimensional (3D) cell culture to study ovarian cancer[6–10]. Spheroid models of ovarian cancer has been used to evaluate the effectivity of new anticancer agents[11–13].

Developing precision drugs targeting the proliferation of cancer cell is essential for cancer treatment. Natural product compound OSW-1 offers a novel targeting mechanism for cancer cell proliferation. It targets oxysterol-binding protein (OSBP) and OSBP-related protein 4 (ORP4)[14–18]. ORP4 is a known precision target in cancer and essential for cancer cell viability and proliferation[18–20], however, OSBP has no such functionality. OSW-1 has been shown to have a potent cytotoxicity against variety if cancer cell lines in vitro, with an average IC_{50} of 0.78 nM[21].

ORP4 is highly expressed in ovarian cancer according to FANTOM database[22]. Here, we show the cytotoxic activity of ORP-4 targeting OSW-1 in a panel of ovarian cancer cell lines as well as ovarian cancer spheroids. OSW-1 reduced the expression of ORP4 protein both monolayer and spheroids culture and cytotoxic activity of OSW-1 was correlated with ORP4 levels. Removal of lipids in the extracellular environment which is also a feature of limited amounts of exogeneous cholesterol in non-vascularized tumors[23,24], increased the activity of OSW-1. Overall, our results show that natural compound OSW-1 induces cytotoxicity in monolayer and spheroid models of ovarian cancer with a significantly increased effectivity compared to Taxol and Cisplatin and offer a novel path for potential treatment of ovarian cancer.

References

1. Lengyel, E. Ovarian cancer development and metastasis. *American Journal of Pathology* **177**, 1053–1064 (2010).
2. Kipps, E., Tan, D. S. P. & Kaye, S. B. Meeting the challenge of ascites in ovarian cancer: New avenues for therapy and research. *Nature Reviews Cancer* (2013) doi:10.1038/nrc3432.
3. Al Habyan, S., Kalos, C., Szymborski, J. & McCaffrey, L. Multicellular detachment generates metastatic spheroids during intra-abdominal dissemination in epithelial ovarian cancer. *Oncogene* (2018) doi:10.1038/s41388-018-0317-x.
4. Xing, H. *et al.* Knock-down of P-glycoprotein reverses taxol resistance in ovarian cancer multicellular spheroids. *Oncol Rep* (2007) doi:10.3892/or.17.1.117.
5. Gunay, G. *et al.* The effects of size and shape of the ovarian cancer spheroids on the drug resistance and migration. *Gynecologic Oncology* S0090825820338579 (2020) doi:10.1016/j.ygyno.2020.09.002.
6. Pampaloni, F., Reynaud, E. G. & Stelzer, E. H. K. The third dimension bridges the gap between cell culture and live tissue. *Nature Reviews Molecular Cell Biology* vol. 8 839–845 (2007).
7. Sontheimer-Phelps, A., Hassell, B. A. & Ingber, D. E. Modelling cancer in microfluidic human organs-on-chips. *Nature Reviews Cancer* **19**, 65–81 (2019).
8. Paullin, T. *et al.* Spheroid growth in ovarian cancer alters transcriptome responses for stress pathways and epigenetic responses. *PLOS ONE* **12**, e0182930 (2017).
9. Souza, A. G. *et al.* Comparative Assay of 2D and 3D Cell Culture Models: Proliferation, Gene Expression and Anticancer Drug Response. *Current Pharmaceutical Design* **24**, 1689–1694 (2018).

10. Ghosh, S. *et al.* Three-dimensional culture of melanoma cells profoundly affects gene expression profile: A high density oligonucleotide array study. *Journal of Cellular Physiology* **204**, 522–531 (2005).
11. Xing, H. *et al.* Effect of the cyclin-dependent kinases inhibitor p27 on resistance of ovarian cancer multicellular spheroids to anticancer chemotherapy. *Journal of Cancer Research and Clinical Oncology* **131**, 511–519 (2005).
12. Yang, Y. *et al.* Reversing platinum resistance in ovarian cancer multicellular spheroids by targeting Bcl-2. *OncoTargets and Therapy* **12**, 897–906 (2019).
13. Nunes, A. S., Barros, A. S., Costa, E. C., Moreira, A. F. & Correia, I. J. 3D tumor spheroids as in vitro models to mimic in vivo human solid tumors resistance to therapeutic drugs. *Biotechnology and Bioengineering* vol. 116 206–226 (2019).
14. Burgett, A. W. G. *et al.* Natural products reveal cancer cell dependence on oxysterol-binding proteins. *Nature chemical biology* **7**, 639–47 (2011).
15. Roberts, B. L. *et al.* Transient Compound Treatment Induces a Multigenerational Reduction of Oxysterol-Binding Protein (OSBP) Levels and Prophylactic Antiviral Activity. *ACS Chemical Biology*acschembio.8b00984 (2019) doi:10.1021/acschembio.8b00984.
16. Roberts, B. L. *et al.* Differing activities of oxysterol-binding protein (OSBP) targeting anti-viral compounds. *Antiviral Research* **170**, 104548 (2019).
17. Lim, C.-Y. *et al.* ER-lysosome contacts enable cholesterol sensing by mTORC1 and drive aberrant growth signalling in Niemann-Pick type C. *Nature cell biology* (2019) doi:10.1038/s41556-019-0391-5.
18. Zhong, W. *et al.* ORP4L Extracts and Presents PIP2 from Plasma Membrane for PLCβ3 Catalysis: Targeting It Eradicates Leukemia Stem Cells. *Cell Reports* **26**, 2166-2177.e9 (2019).
19. Charman, M., Colbourne, T. R., Pietrangelo, A., Kreplak, L. & Ridgway, N. D. Oxysterol-binding Protein (OSBP)-related Protein 4 (ORP4) Is Essential for Cell Proliferation and Survival. *Journal of Biological Chemistry* **289**, 15705–15717 (2014).
20. Li, S., Ma, H., Zhong, W., Luo, W. & Jiang, Y. OSBP-Related Protein 4L Controls Leukemia Cell Fate Through Oxysterol Signaling to Inositol 1 , 4 , 5-Trisphosphate Receptor 1 , Calcium Fluxes , and Bioenergetics.
21. Mimaki, Y. *et al.* Cholestane glycosides with potent cytostatic activities on various tumor cells from Ornithogalum saundersiae bulbs. *Bioorganic & Medicinal Chemistry Letters* **7**, 633–636 (1997).
22. Uhlén, M. *et al.* Tissue-based map of the human proteome. *Science* **347**, (2015).
23. Daniëls, V. W. *et al.* Cancer cells differentially activate and thrive on de novo lipid synthesis pathways in a low-lipid environment. *PLoS ONE* **9**, 13–19 (2014).
24. Criscuolo, D. *et al.* Cholesterol Homeostasis Modulates Platinum Sensitivity in Human Ovarian Cancer. *Cells* **9**, 828 (2020).

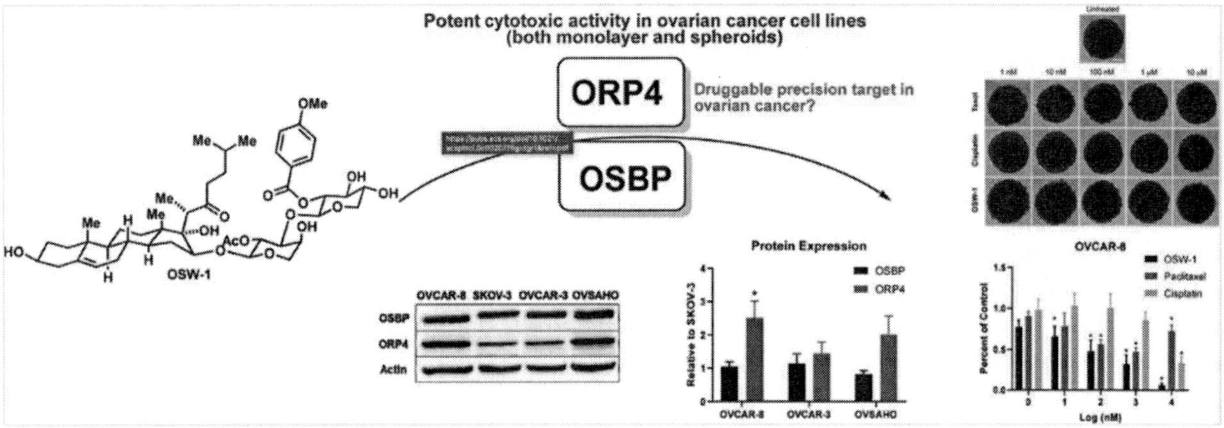

623423 Effect of Signal Processing on Feeding Performance Evaluation

Abstract:

Powder feeding is the start of every continuous pharmaceutical manufacturing line. A constant mass flow into the process is crucial for high quality product. Therefore, feeding performance characterization can aid in detection of feeding issues, even during the early-stage process development phase. Often, many different pre-selected material-equipment pairs with different equipment configurations have to be tested before a suitable setup can be found. Gathering such a large dataset is time and material intensive. A common approach to reduce the experimental in-house effort is to review the data of feeding studies which have been published in literature. Usually, the relative standard deviation (RSD) is used as single measure for the feeding performance comparison. However, signal processing details, like sampling frequency and filter characteristics (cutoff frequency) are rarely reported with the RSD data. However, this additional information is crucial when comparing the results of different feeding studies.

We show in the presentation, how selection of sampling frequency and filter cutoff frequency is affecting the statistical evaluation of feeding performance. Illustrative examples are created to demonstrate aliasing, corrupting the signal dynamics. Furthermore, the effect of sampling frequency on the computed RSD value is highlighted by means of the confidence interval of mean, standard deviation and RSD. Additionally, the RSD is highly affected by the filter characteristics applied to the data. Too conservative filtering might not sufficiently reduce the noise resulting in a too high RSD, whereas too restrictive filtering might remove relevant signal contributions, resulting in a too low RSD. The reliable judgment of feeding performance is not possible in both situations.

We propose, additionally to RSD, to report details about data acquisition and signal processing when evaluating feeding performance based on reference scale data. Furthermore, the confidence interval can give an impression on the statistical significance of the feeding performance characterization, allowing more reliable comparison between the different measurements.

171b – INVESTIGATING THE CORRELATION OF CHEMICAL STABILITY AND PROCESS INDUCED DISORDER: A CASE STUDY OF PHOSPHATE PRODRUG OF VENETOCLAX

Shuang Chen[1], Richard Hong[1], Haichen Nie[1,4], Jie Chen[1], Ahmad Sheikh[1], Moussa Boukerche[1], Yu-Ming Pu[1], Yi-Yin Ku[1], Russell Hertzler[2], Jacquelyn North[2], Lewis Meads[2], Milad Nazari[2], Nathaniel Catron[3], and Sean Garner[3]

1 Process Research and Development, AbbVie Inc., North Chicago, IL, USA
2 Analytical Research and Development, AbbVie Inc., North Chicago, IL, USA
3 Drug Product Development, AbbVie Inc., North Chicago, IL, USA
4 Merck, West Point, PA, USA

Abstract

Venetoclax, the first-in-class selective BCL-2 inhibitor that has transformed the treatment of chronic lymphocytic leukemia (CLL) and acute myelogenous leukemia (AML), is a complex beyond rule-of-five (bRo5) molecule with physicochemical properties that fall well outside those of typical orally bioavailable drugs. This includes high molecular weight, extremely low aqueous solubility and permeability. To improve the drug's bioavailability, ABBV-167, a phosphate prodrug of Venetoclax with better ADME properties was evaluated. The bRo5 BCL-2 prodrug molecule has remarkably increased the solid form complexity and lead to development challenges such as excessively complex speciation, polymorphic landscapes, and significantly altered chemical and physical stabilities. In this study, a comprehensive evaluation was performed to fully assess the crystalline disorders of ABBV-167 hemi-sulfate hexahydrate that are associated to different manufacturing steps including crystallization, drying, and tablet compression. Apart from the traditional evaluation of crystallinity loss using lab X-ray powder diffraction, synchrotron-based high-resolution total scattering pair distribution function (PDF) analysis was also applied to correlate process-induced crystalline disorder with process conditions. The relationship between the degree of crystalline disorders and risk of chemical stability under stressed conditions was elucidated. We found that undesired process-induced crystallinity disorder can significantly influence the chemical stability of prodrug drug substance and further impact the shelf-life and impurity profile of drug product. The study has illustrated the importance and challenges of assessing and controlling process induced crystalline disorders for chemically labile prodrugs with complex solid form landscape.

Disclosure:
AbbVie and Genentech sponsored and funded the study; contributed to the design; participated in collection, analysis, and interpretation of data; and in writing, reviewing, and approval of the final version. All authors are employees or former employees of AbbVie. Haichen Nie is currently an employee of Merck. The authors declare no competing financial interest.

631256 CFD Modeling of Drop Dynamics during Inkjet Based 3D Printing Process

Abstract:

Introduction/Purpose:

3D printing is a rapidly expanding technology in various fields including pharmaceutical industry as it offers multiple advantages such as personalization of medicine, fabrication of complex dosage form and high drug loading. Binder jet 3D printing involves jetting of binder solution on top of the powder layer in layer-by-layer fashion. The properties of the formulation are highly dependent upon jetting of the ink/binder solution from the printhead nozzle. In this study a preliminary finite volume based computational fluid dynamics (CFD) model has been developed to simulate liquid drop formation in a piezoelectric inkjet printhead. Finite-volume based CFD model works by solving continuity and momentum (Navier-stokes) equations to obtain the fluid flow field in the Eulerian coordinate system in conjunction with appropriate physical models in each control volume.

Methods:

In this study Volume of fluid (VOF) and Continuum Surface Force (CSF) model was used to model the drop formation from the printhead nozzle using commercial software ANSYS Fluent. VOF model is used to track interface and CSF model accounts for the surface tension effects. The piezoelectric nozzle simulated here works by creating pressure inside the nozzle when voltage is applied which consequently leads to drop jetting. The system geometry was created in ANSYS Space Claim with a nozzle diameter of 40μm and a stand-off distance of 5mm. Meshing was performed using ANSYS meshing, and model setup was done in ANSYS Fluent. The pressure generation inside the nozzle was implemented by corresponding inlet velocity boundary condition.

Results:

Initially the mesh independence study was performed by analyzing different mesh element sizes and different inlet velocity values to find the optimum mesh element size to avoid the effect of different mesh sizes on the simulation results. Various simulations were performed with different inlet velocities and parameters such as drop diameter, drop velocity and drop shape were compared with the experimental data.

Conclusions:

The drop formation is known to be influenced by the fluid properties such as surface tension, viscosity, and density of the ink solution which were studied using this model for two ink solutions used during the experimentation. This CFD model will help in evaluating the jettability of different ink solutions thereby helping to improve the printing process.

625451 Process Safety Evaluation and Scale-up Resorting to Mechanistic Modelling for Exothermic and Gas Release Reactions

Abstract:

The present work regards all the steps taken during the development and scale-up of an exothermic (Stoessel criticality class 5) and gas releasing (> 1.7m3/kg) reaction, while evaluating the potential impact of the control strategy on the large-scale process conditions/outcome.

Even tough the main reaction is exothermic, the most part of the heat generated throughout the process comes from the decomposition of the reagent in the solvent, with strong gas release. This was determined by performing a blank run without starting material, where it was observed that similar heat and gas releasing profiles were obtained between normal process and blank run. After this finding, it was relatively straight-forward to propose a mechanism that could represent the reaction data such as yield, heat output and gas release, from routine lab experiments and calorimetry studies data.

The following step was to determine the heat transfer capability of the large-scale reactor by performing heating/cooling cycles with the same solvent of the reaction and collecting jacket and reactor temperatures throughout the cycles. Afterwards, it was possible to simulate what-if scenarios at large scale and calculate the accumulated heat through time and determine the best conditions in terms of jacket temperature and feeding duration to optimize yield, productivity and avoiding temperature spikes. We also estimated what would be the outcome in case of cooling failure scenario, which helped to understand and design the safe guards strategy that included how long the cooling could be on hold between changing from the regular heat transfer fluid to liquid nitrogen.

The large-scale campaign was composed of 3 batches of >50 kg (54.4, 54.2 and 94.7 kg) and without any impact on safety (for people and process) nor yield.

625462 Reaction Kinetic Modelling Coupled with Process Safety Calorimetry for Effective Scale-up of an Exothermic Reaction

Abstract:

The present abstract outlines the work carried out for scaling-up an exothermic reaction (ΔT adiabatic = +67°C) from lab scale (10-20g) to industrial GMP scale (more than 50kg), where the foundations of the scale-up where built upon a fit-for-purpose kinetic model, integrated with calorimetry data and large-scale reactor heat transfer characterization. All the models developed were based on the available experimental results. However, due to time constraints, no further experimental data based on the kinetic modelling findings throughout this work could be generated prior to the GMP campaign.

Primarily, to build the initial model, the proposed mechanism was composed of a competitive lumped reaction system (A+B à P, A+B àImp) and it was developed using experimental data gathered during lab development, namely, routine reaction calorimetry studies and exploring different process conditions such as reaction temperature, reagent, and solvent amount. The reaction model was validated at >100g scale experiments.

Once a good fit between experimental data and mechanistic model was obtained, the heat transfer capability of the large-scale reactor had to be determined. This task was performed by heating and cooling cycles with the same solvent of the reaction and spanning a volume around the reactor occupancy at the target batch scale. By collecting jacket and reactor temperatures throughout the cycles it was possible to determine the heat transfer coefficient of the reactor. Having the cooling capacity of the large-scale reactor determined, it was possible to simulate what-if scenarios at large scale, guide operation conditions during execution of 50kg batches and avoid any uncontrolled temperature peaks during reagent addition and throughout reaction duration. Three scenarios were explored, namely reagent addition at 8 and 20°C for a 60kg scale batch and 20°C for a possible future 120kg scale, where the critical aspect was to maintain reaction temperature throughout reagent addition and until reaching the in-process control condition between 5-15°C. The simulated reaction temperature profiles provided to the production engineers were as follows:

(See pictures)

With this information in hand, it was possible not only to better guide the execution of the batches but also to understand which were the most critical tasks during the reaction (e.g., change in jacket temperature, end of reagent feeding and expected reaction final temperature). The execution of 3 GMP batches (60, 160 and 100kg, respectively) were successfully performed and compared well with the proposed simulated profiles.

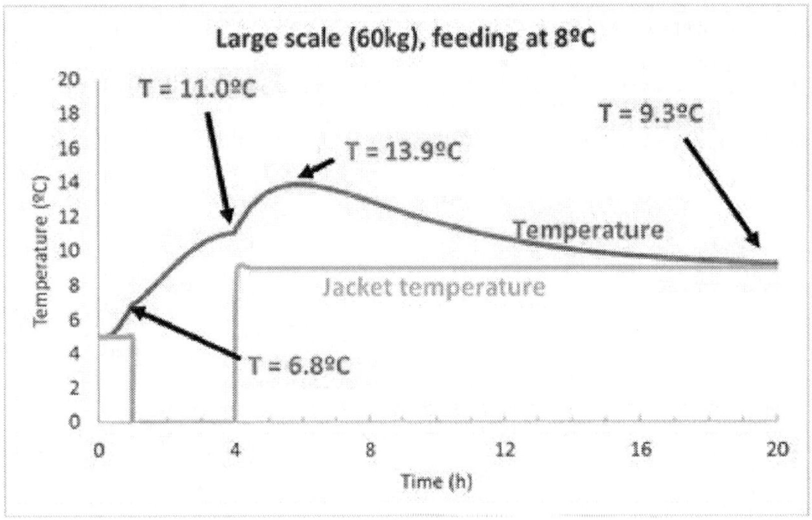
Large scale (60kg), feeding at 8ºC
T = 11.0ºC
T = 13.9ºC
T = 9.3ºC
T = 6.8ºC
Temperature
Jacket temperature
Temperature (ºC)
Time (h)
0
2
4
6
8
10
12
14
16
18
20
0
4
8
12
16
20

630318 Model Aided Scale-up for High Shear Rotor-Stator Wet Milling Process

Abstract:

Particle size specification for an API (Active Pharmaceutical Ingredient) is a Critical Quality Attribute (CQA) that contributes to the bioavailability of the API. The crystallization process is often followed by either wet milling or dry milling. High shear rotor-stator wet milling is used extensively in the industry. However, scaling-up/down of a wet mill is non-trivial[1].

In this work, we demonstrate a new model-aided workflow for Wet Milling scale-up using Population Balance Modelling (PBM) in gFormulate®.The model-aided framework helped us successfully scale-up from the lab-scale (IKA MagicLab <100g API) to the Kilo Lab (IKA 2000/4 - 1.5 kg API) and subsequently to the plant scale (IKA 2000/10 – 40 kg API) with "Right First Time" quality. Traditional high shear wet milling scale-up is based on a single parameter (e.g., Tip Speed, Slot Events[2], Normalized energy[3], & Shear Number[1]). The prediction of milling time based on a single scaling factor hasn't been very successful[2]. The time required for the milling is highly non-linear. Bigger particles break very easily at the start of the milling, however, a large number of batch turnovers are required closer to the grinding limit of the material. A single scale-up parameter wouldn't be able to characterize this behaviour. The model takes into account the detailed geometry of the rotor-stator, the non-linear change of the particle population during milling (breakage kinetics & breakage probability), and the mill residence time.

A colloidal wet mill (IKA MK) was selected over a traditional toothed rotor/stator mill based on the higher shear required to satisfy product particle size specifications close to the grinding limit. Equipment parameters including the number of rotor teeth, stator teeth, & groove size were measured in this work. The breakage kinetics, selection function, & the grinding limit[4] for the PBM were estimated based on the Lab experiments in IKA MagicLab. It has been shown in the literature that the breakage kinetics would scale directly with the Shear Number[1]. The breakage kinetics was directly proportional to the Shear number induced by the equipment geometry. However, Shear Number alone was not sufficient to predict the performance on scale-up from the MagicLab to the KiloLab scale. The model was refined at the Kilo Lab scale to estimate the breakage kinetics coefficient for Shear Number in the Kilo Lab Scale (IKA 2000/4). All other breakage kinetics & function parameters were kept constant. This model was subsequently transferred to the Plant scale (IKA 2000/10) based on the refined coefficient from the Kilo Lab-scale (IKA 2000/4). The model was used to understand the sensitivity of the Key Process Parameters (Gap width, rpm, residence time in the mill) on the milling time & yield.

References:

1. Luciani, C. V., Conder, E. W. & Seibert, K. D. Modeling-Aided Scale-Up of High Shear Rotor–Stator Wet Milling for Pharmaceutical Applications. (2015) doi:10.1021/acs.oprd.5b00066.
2. Harter, A., Schenck, L., Lee, I. & Cote, A. High-Shear Rotor–Stator Wet Milling for Drug Substances: Expanding Capability with Improved Scalability. Org. Process Res. Dev. 17, 1335–1344 (2013).
3. Engstrom, J., Wang, C., Lai, C. & Sweeney, J. Introduction of a new scaling approach for

particle size reduction in toothed rotor-stator wet mills. Int. J. Pharm. 456, 261–268 (2013).
4. Luciani, C. V. Impact of Process Parameters on the Grinding Limit in High-Shear Wet Milling. Org. Process Res. Dev. 22, 1328–1333 (2018).

623418 Development of Process Models Suitable for Controller Design - Use Case: Flow Synthesis of Mesalazine

Abstract:

The implementation of continuous manufacturing routes – both in primary and secondary manufacturing – requires the design of suitable process control concepts. Besides traditional approaches based on standard controllers like proportional-integral-derivative (PID), also model based control concepts, e.g., model predictive control (MPC), are increasingly realized in industry. Such concepts allow the systematic consideration of process constraints and the intuitive formulation of the control objective. However, these model-based approaches require the availability of a suitable process model. The process model needs to describe the real system dynamics at a sufficient accuracy, while keeping the model complexity - in terms of model structure and computational effort needed for performing the predictions - low enough.

In the talk, modeling approaches that are suitable for controller design are presented. Next to mechanistic methods based on mass- & energy-balances and kinetic modeling, also data driven concepts based on the local linear model tree (LOLIMOT) algorithm are outlined. The former one offers more process insight, but might be infeasible for complex processing steps where the process understanding is insufficient to apply the mechanistic approaches. The latter one allows the modelling of reactions where the underlying mechanisms are not known, but a sufficient amount of experimental data is available. The concepts are demonstrated via the use case of continuous Mesalazine synthesis, which consists of three main steps: nitration, substitution and hydrogenation. After each of the steps, PAT devices are installed to measure the respective reaction outcome. The captured data is used to parameterize the models and to test the model accuracy.

628310 Continuum Simulations of Multiphase Heat and Mass Transfer in Wet Granular Mixtures Modeled As Separated Phases

Abstract:

Heat and mass transfer considerations in multiphase flows such as wet granular mixtures, where solid, liquid, and gas co-exist, are important in many industrial processes such as pharmaceuticals manufacturing and petrochemicals processing. They are also paramount in key industrial processes involving solvent removal by thermal drying of powders and wet granules. Continuum simulations involve solving the averaged equations for conservation of mass, momentum, and energy by treating the granular medium as a continuum. The interphase exchange terms are equal but of opposite signs in the phases, and accurate solutions of the coupled set of equations require solving them in terms of transformed variables that span a subspace of reduced dimension. This is because the singular value decomposition of the matrix corresponding to the coupled set of equations shows it is singular. In this study, we propose a novel iterative solution procedure to solve the coupled heat diffusion equation in three phases if one or more singular values of the matrix are zero. Singular value decomposition is used to solve the matrix in a reduced subspace by eliminating the zero singular values, and the corresponding columns of the left and right singular vectors. Continuum multiphase simulations of heat transfer between solid, liquid, and gas phases, and mass transfer between the liquid and gas phases in a stationary granular bed, are performed using the proposed iterative procedure. Further realities of particle drying due to solvent boiling and wet bulb evaporation mechanisms are incorporated into the coupled heat and mass transfer Computational Fluid Dynamics (CFD) simulation to obtain realistic results. The numerical results obtained using the multiphase continuum simulations are verified using a previously developed analytical model for heat transfer in a cylinder filled with stationary wet granular material, to establish accuracy.

633176 2D-Inkjet Printing for Pharmaceutical Particle Engineering

Abstract:

Modern, highly potent API's often show poor solubility and handling characteristics. Spray drying has been used as a Particle Engineering platform enabling the simultaneous improvement of the biovailability and processability of API's. As spray drying has several drawbacks, such as inconsistent quality (owing to the challenges in the control of the droplets' size), difficult scale up and the necessity of a large feedstock, the development of alternatives is of high potential for the (bio) pharmaceutical sector. The possibility of using 2D-inkjet printing as a replacement to such spray drying processes was investigated. By dispensing a controlled amount of liquid on a substrate, consistent quality can be achieved. The production can be easily scaled out by numbering up the nozzles. Also only a very limited amount, of a few grams of material, is needed for initial feasibility studies.

In order to reproducibly produce droplets and particles, a stable formulation needs to be found. Also the re-crystallization and drying kinetics are of high importance, as they influence the properties of the final particle severely. Formulations consisting of surrogate substances, API's, polymers and other excipients were tested and analyzed regarding their stability. The surface/droplet interactions were optimized by altering the substrate material and surface roughness, in combination with the volume of the deposited droplet. Dimensionless numbers provided the guidance for this activity. Finally, the drying and solidification were manipulated by altering the process temperature.

By identifying the critical parameters for 2D-inkjet printing process it was possible to achieve spherical particles, starting from very poor solidification behavior (i.e. coffee ring formation, flat structures). These particles were further characterized regarding their polymorphic forms (DSC, SWAXS) and shape (SEM), to find the optimal process for the used substances.

Summarizing, this presentation will show how to use 2D-inkjet printing for particle engineering purposes, leading to reliable and reproducible results for pharmaceutical applications.

627077 Innovations in Microbiology Imaging: Automated Gram Scanner

Abstract:

Background: The Gram stain is an essential test of pharmaceutical microbiology. With its swift and simple execution, the Gram stain serves as the first critical step in providing results to researchers regarding biocontamination during drug manufacturing. Even with escalating expenditures and a dwindling workforce, laboratories are still expected to meet this enhanced demand for rapid and accurate test results. Conversely, utilizing standard manual microscopy to assess Gram-stained slides is time-intensive, tedious work. This has led to the need for a more ergonomic and cost-effective solution of an automated microbial workflow.

To meet this challenge, we developed the Gram Scanner (GSC), an automated microscope scanning and imaging system. The aim of this study was to evaluate the amount of technician time as well as laboratory costs that the GSC can save a pharmaceutical microbiology lab. We hypothesized that automated scanning and image acquisition of gram-stained slides would be a faster process compared to the standard manual microscopy technique.

Methods: Gram-stained blood slides were scanned automatically with the Metafer Gram Scanner (MetaSystems Group Inc, Newton, MA) as well as evaluated manually through the oculars of a Zeiss AxioImager Z2 (Zeiss AS, Norway). For the manual workflow, 10 fields of view (FOVs) of each slide were first viewed under a 10X objective (pre-scan), and then each slide was analyzed under a 100X oil immersion objective (OIO), viewing 10, 20, 40 and 50 FOVs. Slide results were entered into Excel to mimic LIS data entry. For the automated workflow, 50 FOVs of each slide were first captured for the 10X pre-scan and then each slide was scanned under a 63X OIO, capturing 10, 20, 40, and 50 FOVs. Slides were evaluated by viewing the digitized images captured in the Metafer gallery, and results were entered into the scanning platform's reporting tool (Neon). To determine the financial benefits of the GSC, we developed a return on investment (ROI) economic model based on known average wages and fringe benefit rates for medical technologists, the average number of slides processed daily in microbiology laboratories, and equipment cost including a service contract.

Results: Our overall findings confirmed that the GSC lessens the time as well as the financial burden for pharmaceutical microbiology laboratories. By reducing the amount of time that technologists are tethered to the microscope, the GSC saved ~85% of tech time in comparison to the manual method. When separating the automated workflow time into the time that a technologist is required to physically interact with the microscope or computer (hands-on time) and the time in which the machinery is independently performing a task (automation time), we found that the automated workflow was significantly comprised of only 17-20% hands-on time in comparison to 100% of hands-on time comprising the manual workflow. Additionally, linear regression analysis indicated that the amount of time for a technologist to review and score images generated by automated scanning only increased by 0.6 seconds with each 10 additional FOVs, while the review time increased 3.5 seconds for every additional 10 FOVs for slides reviewed manually. Lastly, ROI analysis revealed an estimated cost savings over 5 years for labs scanning an average of 3,000 slides monthly was over $1 million.

Conclusions: Using the GSC, pharmaceutical microbiologists can reduce microscope time significantly, increase overall lab efficiency through more cost-effective use of technician time, and thus aid in quickening the drug manufacturing process. With a breakeven in 1.3 years, the GSC is an ergonomic investment in full lab automation that permits labs to scale up production without increasing their labor force.

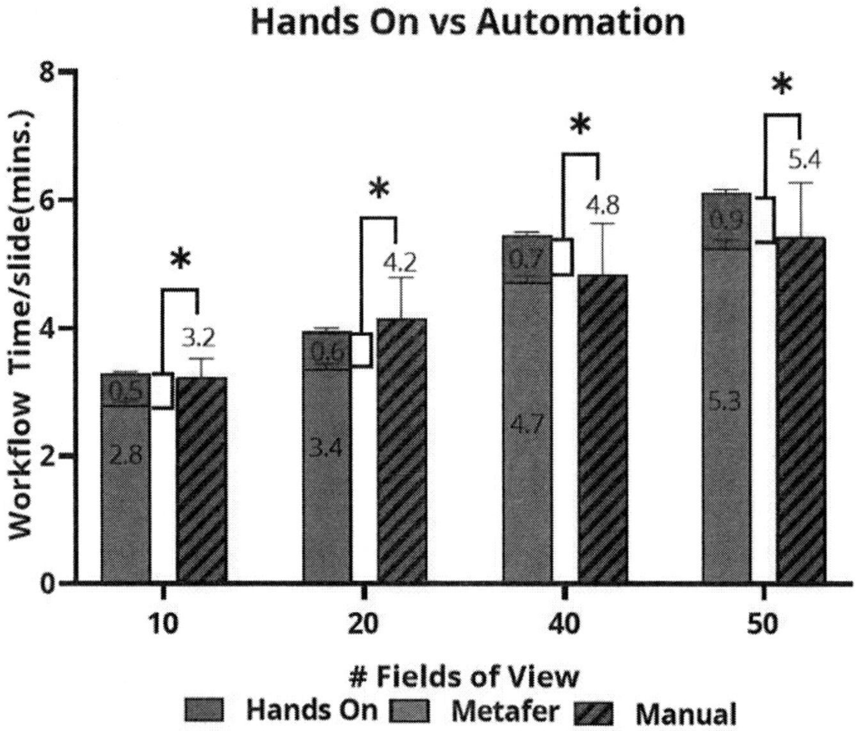

Figure 1. Automation through Metafer results in significantly less time compared to the manual method in which technologists are required to be at the microscope or computer. Data expressed as ± SD. Statistical significance, * $P<0.5$, determined by two-way ANOVA with Tukey's post-hoc test, n=5 slides per group.

622682 Implementation of Model-Based Control Strategies in Manufacturing of Drug Substance and Drug Product

Abstract:

During the last years the shift from batch to continuous manufacturing was accompanied by extensive research in development and testing of advanced control concepts based on in-line process data and process models. A maturity level of these methods is now reached to enable transfer to industrial pharmaceutical processes. Still the majority of commercial manufacturing lines remain at an automation level, at which end product testing, parametric control or process execution within strictly defined spaces are applied to maintain product quality and a state of control. Emphasis is now required, especially for new drug products, to make use of the aforementioned alternative approaches.

This talk will give a step-wise overview of the development and implementation of model-based strategies for process automation. It will include the PAT concept, the control strategy, process model development and controller synthesis. Examples will be shown for different manufacturing processes. Process automation for start-up, steady state operation and exceptional events handling will be discussed for an API synthesis and a solid dosage production process. In the first one, mesalazine is continuously synthesized in flow via nitration, separation, substitution and hydrogenation. In the second one, tablets are produced via pelletization of hot melt extruded strands. The talk will outline opportunities and challenges of the model-based automation approach.

630650 Developing Predictive Models for Degradation of APIs during Hot Melt Extrusion in Pharmaceutical Processes

Abstract:

Drug product manufacture requires a demonstrated control of all parameters which may impact product quality. While several common methods exist to manufacture a drug product, Hot Melt Extrusion (HME) has gained wide acceptance to produce amorphous solid dispersions with high bioavailability and consistent, sustained release;[1] however, the high temperatures of the melt during extrusion can lead to decomposition of the Active Pharmaceutical Ingredient (API). Developing a thorough understanding of the fundamental scientific phenomena that govern degradation of an API during the conditions of HME helps to provide a robust control strategy and a well-defined operating space. Stability of the API under the conditions of HME is a function of many variables, including heat transfer, mass transfer, thermodynamics, physical properties of the formulation, impurities in the matrix, and intrinsic reaction kinetics. To fully map out the experimental space to account for all of these factors can lead to significant time and material constraints, especially during early stages of development, wherein multi-kg quantities of API may not yet be readily available. For this reason, demonstration of appropriate scale-down methodologies to evaluate such degradation adds significant value to the drug product development lifecycle.

In this presentation, we will provide a methodology for elucidating the degradation mechanisms and governing factors that control API decomposition to form impurities during HME. Such a methodology begins through generation of 1D models to predict the temperature profile that results from selection of the extruder operating conditions and by heat generation from viscous dissipation.[2] Further, the important role of screw configuration and operating conditions on the residence time distribution of the melt will be discussed. Selection of an appropriate solvent system can then be used to mimic the solvation environment and impurities of common pharmaceutical excipients to deconvolute the intrinsic reaction kinetics. By combining the residence time distribution and temperature profile with the intrinsic reaction kinetics, we will discuss how we have applied this understanding to model API degradation to de-risk the operating space for manufacture of drug products by HME.

[1]. Repka, M. A., et al., *Drug Dev. Ind. Pharm.*, 33 (2007) 1043-1057.

[2]. Eitzlmayr, A., et al., *Int. J. Pharm.*, 474 (2014) 157-176.

638756 All vials are not the same: Potential role of vaccine quality in vaccine adverse reactions

Abstract:

Pharmaceutical manufacturing is not very precise. It is unrealistic to expect zero product defect rate. This presentation discusses limitations of current quality control practice for vaccines, the potential role of vaccine quality in vaccine adverse reactions, and an emerging analytical technology to enhance quality control for vaccines and other biologics.

627516 Enabling Agile Process Development and Novel Chemistries through Flow Processes

Abstract:

Abstract: For the last two decades continuous manufacturing (CM) and flow chemistry have been constant topics of conversation in the pharmaceutical industry. Yet, despite environmental, economic, and time-to-market benefits, the adoption of CM technologies still faces resistance. Reasons for this include a lack of experience implementing CM processes, a natural propensity to stick with the status quo, build-up of appropriate infrastructure from lab to plant, and concerns around filing strategies. In a partnership between our Process R&D and Environment Health and Safety we are investigating technology and workflows to address the first two limiting factors. To address these constraints, we have developed a strategy for evaluating, developing, and applying CM technologies across our small molecules and antibody-drug conjugate pipelines. In cooperation with external vendors, we are developing a walk-up flow chemistry bench equipped with a variety of fluid dosing, reaction, and Process Analytical Technologies (PATs) that will enable chemists from across the company to easily evaluate multi-step continuous processes. The system will critically enable automated design space generation for flow processes, while also giving users the flexibility to interface with enabling technologies including photochemical and electrochemical synthesis. By combining advanced automation with data-driven experimentation, kinetic modeling, Artificial Intelligence (AI) process modeling and an intuitive yet robust interface, we plan to make flow chemistry process R&D accessible and harmonized within our organization.

Disclosures: All authors are employees of AbbVie and may own AbbVie stock. AbbVie sponsored and funded the study; contributed to the design; participated in the collection, analysis, and interpretation of data, and in writing, reviewing, and approval of the final publication.

629723 A Digital Twin for the L.B. Bohle Tablet Coater

Abstract:

The coating of pharmaceutical tablets to achieve desired properties is accomplished via a common film coating unit operation. Typically, this unit operation is achieved by coating tablets in a rotating drum, which is equipped with multiple nozzles that spray this coating on to the tablets. Along with such spraying operations, the tablets are also exposed to an environment where hot air is blown onto them to dry the obtained coating. Physically, this combination of tablet bed flows, coupled with spray coating and drying represents a complex process, the effectiveness of which is characterized by the inter-tablet coating uniformity as a function of the number of revolutions of the drum. De-risking the performance of the coater at different scales of process development through experimental evaluations is time and resource intensive, although it is of paramount importance. In order to enable a more efficient process design, we have developed a state-of-the-art digital twin of the tablet coating process.

Our digital twin uses a novel first-principles model by combining Computational Fluid Dynamics to simulate spray coating and air flow, and Discrete Element Method to simulate tablet motion. This provides us with unprecedented levels of precision in virtual design, optimization and scale-up, enabling us to study coaters from laboratory scale to commercial scale. Using this digital twin, we have evaluated the effect of tablet loading, drum rotation speeds and tablet types, and we have computed several measures to characterize the performance of the coater. We have then used the result of these models to guide our scale-up decisions. This digital twin is at the forefront of our digitalization initiatives in process development and extends our predictive capabilities, enabling us to reduce developmental time and resources.

621989 Application of Process Modeling and Simulation for Quality Risk Management of Continuous Drug Product Manufacturing

Abstract:

Continuous pharmaceutical manufacturing is an emerging technology that offers production rate flexibility, robustness, and product quality. Understanding process dynamics in relation to material properties (e.g. potency, material flow properties), equipment design (screw types for a feeder), and process conditions (e.g. mass flow rate) allows material traceability throughout the entire manufacturing line. Such understanding is considered essential for identification and mitigation of risks to product quality. Quality risk management is a broad system of considerations and practices, applied across the drug product lifecycle, that encompass the assessment, control, and communication of risks to product quality.

Process models can be used to identify potential risks by estimating the impact of potential variations in the process, equipment conditions, or incoming raw materials on product quality attributes. The control strategy adopted to mitigate the identified risks can be incorporated into the process model, enabling an evaluation of the control strategy's effectiveness in ensuring product quality. Furthermore, process models can provide a formal structure for capturing process knowledge and assumptions, facilitating risk communication among various stakeholders, managing prior knowledge for its use in other product and process development.

In this work, we developed and validated process models for continuous direct compression (CDC) of tablets and demonstrated how these process models can aid the process risk assessment through sensitivity analysis, as well as the assessment of risk mitigation strategies through residence time distribution (RTD) analysis. Specifically, sensitivity analysis is used to quantitatively rank the relative effect of the uncertain inputs on model outputs and identify the process parameters and material attributes that are critical to product quality. The results of sensitivity analysis suggest that the API and the excipients density, their flowrates, the blender's rpm, tablet die fill depth, and main compression force are the most significant factors that highly affect the process operability and tablet quality. Moreover, RTD analysis is used to examine process ability in mitigating the risk of upstream disturbances (i.e., variation in the feeder's flowrate) on product quality. One critical technical challenge was successfully addressed when calculating the propagation of upstream disturbances through the entire process by convolution of the disturbance and RTD. In performing the convolution integral calculation, historical state information is required. However, it is not readily available in process modeling platforms. Thus, a novel approach is taken whereby a partial differential equation is used to propagate historical state information to the current state values. Our validated RTD model served as a tool for material traceability throughout the manufacturing line and facilitated setting the in-process control limits for the final blend composition. The presentation illustrates how these modeling approaches have been used in the regulatory assessment of drug applications employing CDC processes by identifying high risk areas, facilitating the evaluation of risk mitigation strategies, and fostering communications among stakeholders.

Disclaimer:

This abstract reflects the views of the authors and should not be constructed to represent FDA's views or policies.

624380 Crystallization Design to Avoid Liquid-Liquid Phase Separation and Control Drug Substance Physical Properties

Abstract:

Crystallization is a critical purification and separation process used in the pharmaceutical industry. A well-developed crystallization process should not only deliver exceptional chemical purity, high yield and throughput, but also the ability to control physical properties and improve downstream processes (*e.g.*, drug substance filtration and drying and formulation of the drug substance into the drug product). For this case study involving a small molecule drug substance, the challenge of liquid-liquid phase separation (or oiling out) was encountered during process development. Crystallization design changes were required to avoid liquid-liquid phase separation and control residual solvent levels in the final drug substance. The regressed-UNIFAC group contribution theory methodology was used to model the phase behavior and predict liquid-liquid phase boundaries in binary solvent systems including the effects of the drug substance; this data was then used to design a crystallization process that avoided the two-phase liquid region. Process analytical technology (PAT) tools were used to collect *in-situ* data for modeling the crystallization and to guide the selection of optimal process parameters. Crystal morphology and particle size were controlled through seed surface area, control of the crystallization kinetics (nucleation and growth), and application of slurry milling. The combined *in-silico* and *in-situ* techniques from this work provide the basis for understanding and predicting liquid-liquid phase separation during crystallization and can be used to design future processes in complex solvent systems.

627323 Material Sparing Approaches for Predicting Powder Flow Using Machine Learning Methods

Abstract:

The performance of pharmaceutical powders is often determined by the distributions of both shape and size of the constituent particles, but typically they are described by simplistic, and often inappropriate descriptors such as D_{50}. Consequently, most of the information about the materials being characterized is lost and/or misrepresented. Image analysis enables the imaging of all particles within a sample and measurement of multiple particle characteristics resulting in an information rich data set. The complex interdependency between particle size and shape and its combined effect on powder performance metrics such as flowability is a prime problem to be tackled using machine learning techniques.

In this work, we explore the application of unsupervised and supervised machine learning techniques to predict the effect of particle morphology of pharmaceutical powders on their flow function coefficient (FFC). To achieve this, we develop novel feature extraction methods using principal component analysis and convolutional neural networks that capture the particle size and shape distribution of powders and lay foundations for a machine learning framework for aiding material selection in pharmaceutical development. Our results indicate that it is possible to distinguish non-flowing powders (FFC <= 2) from other flow classes (FFC > 2) with high confidence, thereby providing an early indication of poor flow and possible recourse using this material sparing approach. The broad applicability of this approach for predicting other powder performance metrics as well as applicability in other areas of drug development will also be discussed.

631685 In-Depth Mechanistic Understanding of Pibrentasvir Agglomeration to Enable Successful Scale-up

Abstract:

Physical properties of active pharmaceutical ingredients (APIs), such as the particle size distribution (PSD) and extent of agglomeration, often impact downstream drug product processes. Strategies utilized to isolate the API, namely the crystallization, filtration, and drying unit operations, must be designed to ensure consistency in the resultant API physical properties. Development and deployment of non-traditional isolation strategies, for an API prone to agglomerate formation, is often necessary to ensure that the API meets the acceptance criteria for PSD.

In this presentation, we will describe the strategies utilized to develop the API isolation for pibrentasvir, a non-nucleoside inhibitor of the hepatitis C virus NS5A replicon. While the pibrentasvir crystallization consistently yielded non-agglomerated material, drying in the agitated filter dryer resulted in agglomerates on the order of 10 to several 100 µm, with varying degree of agglomerate strength as a result of agitation in the presence of residual solvent. In this presentation, we will discuss a mechanistic understanding of the parameters that impact the extent of agglomeration of pibrentasvir and methods for quantifying the strength of the resulting agglomerates. To address agglomeration, the use of non-traditional isolation strategies were employed to ensure significant removal of the residual solvent and prevent the formation of agglomerates through crystalline bridges. This work demonstrates a systematic approach for understanding the formation and characterization of API agglomerates and provides a novel approach to address agglomeration in an agitated filter dryer. Successful deployment of the isolation strategy enabled scale-up of the drying process to yield API that met the acceptance criteria for PSD.

All authors are employees of AbbVie and may own AbbVie stock. AbbVie and Enanta sponsored and funded the study; contributed to the design; participated in the collection, analysis, and interpretation of data, and in writing, reviewing, and approval of the final publication.

623454 Influence of the Rheology of Lactose Excipients on the Feeding Consistency in Tabletting Processes

Abstract:

Minimizing the variability in the feeding process is a crucial challenge in oral delivery systems (e.g. tabletting, capsule filling) to ensure good quality and consistency of the end product. To control and reduce the variability in the feeding rate, it is important to characterize the material properties and bulk behaviour of the powder. It is known that good flowing properties of pharmaceutical powders are required to achieve uniform flow into the die cavities. Insufficient flow can lead to uneven filling of the dies, resulting in large weight and dosage variations of the final tablets. Moreover, at high tableting speeds, the time to fill dies is reduced, making flow properties of the blend even more important.

Lactose is one of the most widely used excipients in the pharmaceutical industry. There are many reasons for its popularity, such as the fact that lactose is largely inert, relatively inexpensive, safe, available in many different grades, and has a long history of usage in successful formulations world-wide. For direct compression processes like tableting, lactose excipients can be used as a filler-binder to provide bulk density, compaction, and flow to the formulation.

In this study, the rheological properties of lactose powders have been investigated with the rotating drum method (GranuDrum) and correlated with the actual consistency of flow achieved in a tableting system. The investigation of powder flowability at different drum rotating speeds gives useful information on the evolution of powder processability at higher shear rate. Indeed, with the historical repose angle characterization method, the stress state at which the powder is submitted is far from the conditions encountered in a tableting machine. In this work, we observed that some excipient grades exhibit shear-thinning behaviour (i.e. an increase of flowability with increasing applied stress) while others show the opposite (i.e. a shear-thickening behaviour). The classification of powder cohesiveness is therefore inverted at high shear. These observations have been correlated with the mass variation of tablets produced with a RoTab tableting press at different feeder speeds. We present the correlation between the flowability assessments obtained with the rotating drum and the weight consistency of the produced tablets. We thus highlight the importance of powder characterization at a stress state equivalent to those at which the powder is processed.

626104 Characterizing the Effect of Particle Properties and Surface Topography Towards Pharmaceutical Powder and Surface Adhesion Using the Enhanced Centrifuge Method

Abstract:

The behavior of pharmaceutical powders during processing is impacted by the adhesion of the individual particles which comprise the powder. Problems such as poor flowability, dust hazards, and equipment wear arise due to uncontrolled particle adhesion and can lead to production challenges. This adhesion behavior, in turn, is critically influenced by the complementarity between the topography of a surface and the shape and topography of the particles that adhere to that surface. Computational models have been developed to predict the behavior of highly idealized powders (i.e. powders comprised of simple geometries such as spheres) under various processes but are limited in their ability to model and optimize the manufacturing and handling of powders comprised of many complex particles. This work focuses on quantifying the real particle property effects of particle size, shape, and topography of three pharmaceutical powders towards the adhesion of three pharmaceutical powders to stainless steel surfaces through a novel experimental and modeling framework called the enhanced centrifuge method that maps particle and surface scale properties onto experimentally-validated adhesion parameters. The result of this work is a validated semi-empirical model for the powder adhesion which describes the behavior of the full powder, across the full range of particle sizes and shapes, against the steel. This model is presented in terms of van der Waals forces, although the approach can be applied to describe capillary-, electrostatic-, or deformation-driven adhesion.

Disclosures:

Andrew Vogt, Maxx Capece and William Ketterhagen are employees of AbbVie and may own AbbVie stock. AbbVie and Purdue contributed to the design; participated in the collection, analysis and interpretation of data, and in writing, reviewing and approval of the final presentation.

627255 Surfactant/Preservative Instability in Injectable Pharmaceutical Formulation

Abstract:

Recent study of incompatibility between surfactant molecules and antimicrobial agents has caused concerns for future protein formulation of multi-dose therapeutic drugs in the pharmaceutical industry. In particular, polysorbate 80 (PS80), a nonionic surfactant widely used in formulation, has been found to be incompatible with m-cresol, a commonly used antimicrobial agent. This incompatibility results in increased turbidity caused by unsteady micelle aggregation or coalescence that can progress over weeks. Here, small-angle neutron scattering (SANS) is used to fundamentally understand the structural origins of this incompatibility. We observed that PS80/m-cresol solution instability is influenced by storage temperature, ionic strength, and component concentration. SANS analysis of PS80/m-cresol solutions over a pharmaceutically relevant concentration range of each component reveals the cause of aggregation, the coalescence mechanism and aggregate structure. Our results identify two concentration-dependent stability regimes at low (below ≈ 2 mg/mL) and high (5 mg/mL) m-cresol concentrations. In these stable formulations, micelle morphology is highly sensitive to m-cresol concentration. PS80/m-cresol solutions aggregate at m-cresol concentration between the two stable regimes. SANS results reveal the aggregation mechanism, and kinetic modelling of the results provides a simplified empirical equation capable of describing and predicting the multiple aggregation phases observed in these solutions. This work aims to develop insight into surfactant/preservative stability for pharmaceutical formulation, which can be leveraged to better understand the effect of excipients on surfactant self-assembly.

625342 Deagglomeration By Conical Milling of a Temperature Sensitive Amorphous Peptide

Abstract:

Conical milling is a common pharmaceutical unit operation in both drug substance and drug product processing. Milling is often implemented after granulation and drying operations as a method of particle size control or more simply as a pharmaceutical elegance operation. For traditional small molecule synthetic active pharmaceutical ingredients (APIs), basic milling process design criteria are well understood and conical mills can be implemented with limited-to-no development. For APIs that are amorphous, temperature sensitive, or both, additional understanding of the material properties and operating conditions are necessary. Characterization of second order thermal transitions of amorphous APIs are necessary to ensure the material remains a suitable solid without a transition to tacky material. Dynamic mechanical analysis and differential scanning calorimetry are used to assess the temperature depended physical property risks to the API during milling. An understanding of the API stability and the thermal profile of the conical mill along with the material characterization gives the necessary information for a rationale design of the milling operation. Applying this methodology to Compound A, an amorphous peptide, a direct scaling of the process from 20 g to 100 kg was successful, while ensuring robust control to deliver high purity API.

625283 Developing Quantitative Chemometric Models for Monitoring API Disproportionation Process with Raman Spectroscopy and X-Ray Diffraction

Abstract:

For API molecules that follow the "Spring and Parachute" model as a tablet goes through the stomach and gastrointestinal tract, the primary objective of tablet formulation is typically to mediate the sudden decrease of solubility by maximizing the hang-time. For development compound X, a formulation ingredient intended to manipulate the microenvironmental pH of the tablet triggered an out-of-proportion enhancement in performance. Investigation of the solid-state property of the freebase-salt continuum revealed a unique co-crystal which forms unexpectedly within the formulation. The co-crystal formation dramatically alters the path of the salt break-down, and subsequently contributes to a significant delay in the salt-to-freebase conversion.

X-ray diffraction and Raman spectroscopy were utilized to monitor the disproportionation process of the API in aqueous media in an in-vitro setting. Process samples (slurry) periodically taken from the vessel were filtered to enable X-ray analysis, while Raman provided continuous spectroscopic tracking of the API suspension in the system. Due to severe peak overlapping, the presence of multiple freebase polymorphs presented a tremendous hurdle to X-ray and Raman data analysis, limiting initial analysis to be (predominantly) qualitative in nature.

The quantitative aspect of the study yielded details of the mechanisms by which the API molecule trans-configure through multiple stages of solid state entities (salt, co-crystal, freebase polymorph 1, 2, 3...) as pH evolves in the GI tract. Mathematical algorithms were developed to address the challenges of quantitative analysis of spectroscopic datasets. A deconvolution algorithm based on random projection of powder X-ray diffractograms was developed to quantify the fraction of freebase polymorphs present in the process. A partial least square model was then built to relate the freebase fractions predicted from X-ray data to Raman spectroscopic data collected in situ. The partial least squares algorithm successfully modeled the covariance structure between Raman spectra and freebase fractions in the presence of severe peak overlapping. Together, these quantitative methods allowed for accurate and high-resolution experimental monitoring of the salt disproportionation process. More importantly, the mathematical methodologies presented in this talk provides a generalized framework that can be broadly applied to any system with a complex polymorph landscape involving spectroscopic data with overlapping peaks.

624669 Prediction of Powder Flow of Pharmaceutical Materials from Physical Particle Properties Using Machine Learning

Abstract:

Background

Understanding powder flow and how it affects pharmaceutical manufacturing process performance remains a significant challenge for industry, adding cost and time to the development of robust production routes. It is therefore crucial improve our understanding of powder flow and how this relates in particular to both tableting and capsule filling. Tablet formation can be addressed by several techniques; however, direct compression (DC) and wet granulation (WG) are the most widely used in industrial operations. DC offers lower-cost manufacturing as a consequence of the fact that it provides streamlined process with fewer steps compared with other unit operations. However, to achieve the full potential benefits of DC for tablet manufacture, this places strict demands on material flow properties, blend uniformity, compactability, and lubrication, that need to be satisfied. On the other hand, WG improves flowability and compactability while preventing segregation through the use of binding agents and secondary process steps although it is not suitable for materials which are sensitive to heat or moisture. WG is, therefore, more expensive and time consuming. Due to these economic advantages, DC is increasingly the preferred technique for pharmaceutical companies for oral solid dose manufacture, consequently making the flow prediction of pharmaceutical materials of increasing importance [1,2]. Flow properties are influenced by the particle size and shape, which are defined during crystallization and/or milling processes. Currently suitability of raw materials and/or formulated blends for DC requires detailed characterization of the bulk properties. A key goal of digital design within Industry 4.0 is to be able to better predict properties whilst minimizing the amount of material required and time to inform process selection during early stage development.

This work aims to improve decision making for manufacturing route selection in the early stages of the drug product manufacturing process development when time and amount and cost of available material are at a premium. A Machine Learning (ML) model approach is proposed to predict the flow properties, specifically the flow function coefficient (FFc) of a new material from its physical particle properties. The aim is that the model produced as an outcome can help decide whether a new material is suited for DC.

Methods

To build a ML model, and hence, to make reliable predictions, a large data set is necessary. Particle size, shape distribution, and powder FFc of more than 100 materials were collected as training data to the supervised and unsupervised ML algorithms used. Particle size and shape distribution were determined using dynamic image analysis (QIC/PIC – Sympatec). This equipment is appropriate for materials in a range of sizes between 30 μm to 8,665 μm, and it only requires less than 2 grams. A Freeman FT4 (Freeman Technology, Malvern, UK) was used to carry out a shear cell test to measure the FFc. The consolidation stress and pre-shearing normal stress in the experiments was 9 kPa. Normal stress for shearing at 7 kPa, 6 kPa, 5 kPa, 4 kPa, and 3 kPa were used, and the sample is sheared to obtain five yield points at different normal and shear stress. Triplicates of each sample were measured using the 25 mm x 10 mil split-vessel.

These materials were classified into three classes regarding their FFc: cohesive, easy-flowing and free-flowing materials belong to class 1 (FFc < 4), class 2 (4 < FFc < 10), to class 3 (FFc > 10) respectively[3]. Generally, for the materials to be suitable for DC, they should be free flowing (class 3). A Principal Component Analysis (PCA), as an unsupervised ML algorithm, was carried out on the data as a dimensional reduction analysis. The ML algorithms selected for classification problems were Random Forest (RF), Neural Network (NN), Logistic Regression (LR), Support Vector Machine (SVM), k-Nearest Neighbor (kNN), Naïve Bayes (NB), and AdaBoost (AB). Furthermore, they were evaluated by 10-fold cross-validation and with an external data set. For the external validation, 9 materials were left out of the data set used to build the model; 3 of them were cohesive (class 1), 3 were easy flowing (class 2) and the remaining 3 materials were free flowing (class 3). Only particle size and shape measurements were fed into the model to validate it.

Results

The data set used to build the ML model included 115 fully characterised materials. Regarding particle size distribution, 42 of the materials had a D_{50} smaller than 100 μm, 56 had a D_{50} between 100 μm and 250 μm, and 17 had a value of D_{50} greater than 250 μm. Concerning particle shape, most of the materials' aspect ratio falls between 0.6 and 0.8, which corresponds with a cubic shape [4]. Based on the FFc, the 27 materials were classified into class 1 (cohesive materials with FFc < 4), 34 were classified into class 2 (easy-flowing materials, $4 <$ FFc < 10), and 54 materials were classified into class 3 (free-flowing materials, FFc > 10). The physical particle properties were ranked according to their correlation with the target variable, resulting particle size distribution (D_{10} and D_{50}) and aspect ratio distribution (a90) the variables that provided more information.

Among the models selected, RF exhibited the best performance statistics, with a value of Area under ROC of 0.807, and a value of classification accuracy of 0.67. Regarding the external validation, AB was the algorithm that could predict the class of more materials correctly (8 out of 9), followed by RF and kNN, that predicted the class of 6 materials correctly.

Implications

The ML model's implementation enables the prediction of the material flow properties (FFc) from size and shape enabling rapid decision-making regarding manufacturing route selection. This could of course be extended to inform formulation optimization or even to provide a performance target for particle engineering efforts. The accuracy and precision of the model is reasonable; however, it can be improved with the availability of more data from a wider training set of materials. This work suggests that particle size and shape distribution are sufficient to enable prediction of flow properties. Particle size distribution, particularly D_{10}, is the variable that has the biggest impact.

References

1. Schaller, B. E., Moroney, K. M., Castro-Dominguez, B., Cronin, P., Belen-Girona, J., Ruane, P., ... & Walker, G. M. (2019). Systematic development of a high dosage formulation to enable direct compression of a poorly flowing API: A case study. *International journal of pharmaceutics*, *566*, 615-630.
2. Trementozzi, A. N., Leung, C. Y., Osei-Yeboah, F., Irdam, E., Lin, Y., MacPhee, J. M., ... & Zawaneh, P. N. (2017). Engineered particles demonstrate improved flow properties at elevated drug loadings for direct compression manufacturing. *International journal of pharmaceutics*, *523*(1), 133-141.
3. Zegzulka, J., Gelnar, D., Jezerska, L., Prokes, R., & Rozbroj, J. (2020). Characterization and flowability methods for metal powders. *Scientific Reports*, *10*(1), 1-19.
4. Yu, W., Liao, L., Bharadwaj, R., & Hancock, B. C. (2017). What is the "typical" particle shape of active pharmaceutical ingredients?. *Powder Technology*, *313*, 1-8.

A Machine Learning (ML) approach was proposed to optimize the manufacturing-route selection from the physical particle properties of a pharmaceutical material.

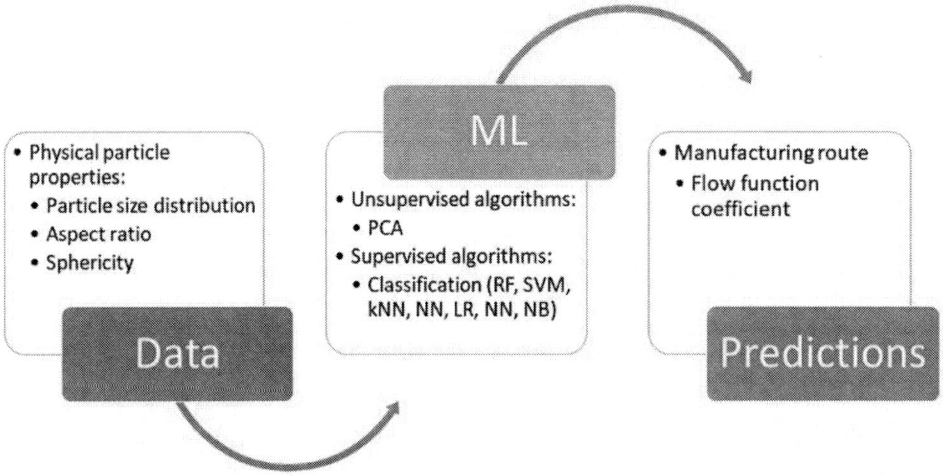

621512 A Systematic Approach to Understand Impurity Incorporation during the Crystallization of Pharmaceutical Drug Substance

Abstract:

Impurity generation during reactions is an inevitable and undesirable event in pharmaceutical drug substance synthesis, which is often tackled by introducing isolation methods downstream. Workups like distillation and liquid-liquid extraction are not well suited for rejection of related substance impurities that do not partition during extraction, and typically crystallization is the method of choice for impurity rejection. However, uncontrolled, or an ill-designed crystallization process may not reject impurities within the specification limits. While process improvements exist for overcoming this challenge, a streamlined approach which involves identification of the impurity incorporation mechanism prior to designing the rejection process provides mechanistic understanding for the development of a robust control strategy. Utilizing such an approach at the time of initial process design can significantly decrease the time required to design a controlled process and reduce the reliance on spiking studies for impurity control.

In this talk, we present a bottom-up approach for impurity control based on a previously demonstrated workflow for understanding impurity incorporation in organic synthetic molecules. [1] Through a case study involving an example organic molecule and its derivative impurity generated during a relevant synthesis process, we uncover the impurity incorporation mechanism. The novel experimental design presented limits the amount of input material required, making this analysis feasible even in scenarios with tight material constraints, which is typically the case in most pharmaceutical process development efforts. Through isothermal equilibration, agglomeration assessment, and slurry washing experiments, we explore solubility-limited impurity retention as well as physical incorporation during crystallization. Lastly, we present a scenario in which unintended reactions can occur during antisolvent crystallization that result in impurity generation and retention. We suggest ways to track the formation and purge of impurities in such a case. Ultimately, we make suggestions for process improvement to achieve a higher degree of impurity rejection and show the effect of their implementation on the API-impurity system discussed.

References

1. Urwin, S.J., et al., *A structured approach to cope with impurities during industrial crystallization development.* 2020. **24**(8): p. 1443-1456.

621905 Structured Approach to Optimize Pharmaceutical Crystallization Process for Optimal Impurity Rejection

Abstract:

Current pharmaceutical drug substance or API manufacturing processes routinely face the challenge of having to control numerous organic impurities simultaneously while also trying to optimize yield. In this talk, we will present an industrial case study for an API in development where we used a structured approach described recently [1] to identify the impurity rejection mechanism of several key impurities present in the process with the aim to support the building of the impurity control strategy. This more mechanistic rejection information is then used to help assess risks of the process and define and optimize the final isolation conditions. A self-consistent approach to data reconciliation from isothermal equilibration, crystallization hold studies and stepwise dissolution experiments will be discussed to aid in the mechanism determination.[2, 3] Based on the solubility limited rejection mechanism identified for several impurities we modified the crystallization process for optimal impurity rejection and process yield as a function of various process parameters using existing mass balance equations.[4, 5]

References

1. Urwin, S.J., et al., *A structured approach to cope with impurities during industrial crystallization development.* 2020. **24**(8): p. 1443-1456.
2. Myerson, A., *Handbook of industrial crystallization.* 2002: Butterworth-Heinemann.
3. Moynihan, H.A. and D.J.R.a. Armstrong, *Stepwise dissolution and composition determination of samples of multiple crystals using a dissolution medium containing aqueous alcohol and fluorocarbon phases.* 2019. **9**(37): p. 21405-21417.
4. Nordstrom, F.L., et al., *Solubility-limited impurity purge in crystallization.* J Crystal Growth Design, 2019. **19**(2): p. 1336-1346.
5. Teerakapibal, R., et al., *Material Impurity Distribution of Lattice-Incorporated Impurities in Salicylic Acid.* J Crystal Growth Design, 2020. **20**(3): p. 1716-1728.

621788 Advances in Crystal Structure Prediction for Conformationally Flexible Molecules

Abstract:

Crystal structure prediction of active pharmaceutical ingredients and other real-world species is frequently complicated by conformational flexibility. One particular challenge stems from delocalization error in commonly used density functionals, which can systematically bias predictions in certain classes of conformational polymorphs. Combining periodic density functional theory with intramolecular conformational energy corrections computed at a higher level of theory can substantially improve polymorph energy rankings. A number of examples of this problematic behavior and how it can be overcome will be presented, with applications ranging from energy ranking select groups of known polymorphs to refining entire crystal energy landscapes. In addition, a new, approximate strategy for estimating vibrational free energy contributions will be discussed.

632921 How Much Do Disorder and Entropy Matter in Molecular Systems?

Abstract:

In this talk, I survey our recent work using molecular dynamics to examine the free energy differences between enantiotropic crystalline systems, where the approximation that the lowest lattice energy crystal is the most stable must necessarily break down. We have found that although small rigid crystals are well-described by quasiharmonic approximations, even moderately conformationally flexible molecules can have extremely complex potential energy surfaces, making it hard to find appropriate structures from which to compute quasiharmonic approximations. We also find that many of these minima are accessible at room temperature and that proper annealing can help identify appropriate ensembles of structures to use for quasiharmonic analysis. Finally, we describe new approximations that make it easier to estimate the free energy difference between crystal polymorphs, including entropic effects, at a level that may be useful for large-scale screening of potential crystal structures.

625794 Immobilized Transmembrane Kinase Receptors As Screening Targets for Neurological Disorders

Abstract:

The treatment of neurodegenerative diseases remains as a great medical challenge, which affects millions of people. Currently, no truly effective treatments for neurodegenerative diseases are available. The failure of clinical translation of nearly all (> 400) tested compounds for Alzheimer's Diseases (AD) triggered interested to develop alternative approaches of developing neuroprotective therapies to activate and restore neuronal functions. Many studies have shown brain-derived neurotrophic factor (BDNF) and its high affinity receptor, tropomyosin-receptor-kinase B (TrkB), are associated with neurological disorders, such as AD. TrkB is a transmembrane protein that activates a number of biological pathways upon activation. The development of techniques using transmembrane proteins as therapeutic targets is of great importance because over 60% of all modern pharmaceuticals target membrane proteins. Here, a novel drug screening platform based cell-membrane coated magnetic nanoclusters (CMMNs) will be disused, where the functional transmembrane receptors on the coating membrane serve as "smart baits" for direct identification of binding compounds from screening sources and the encapsulated magnetic nanoparticles enable quick isolation of identified compounds.

In this presentation, using TrkB as a model system, we evaluated the potential of this drug-screening platform in identifying binders from an artificial mixture and several plant extracts. The CMMNs were prepared using SH-SY5Y neuroblastoma cells overexpressing TrkB and 200 nm iron oxide nanoclusters were used as core materials. The formation of CMMNs was confirmed with transmission electron microscopy. The presence and activity of TrkB receptors on CMMN surfaces were studied using confocal microscopy with antibody labeling and ATP-ADP conversion experiment. The fishing experiments were performed by incubating CMMNs with an artificial mixture containing a known binder and several non-binders and by incubating CMMNs with *Centella asiatica* (Gotu kola) extracts at 37 °C for 20 minutes. The identified TrkB-bound compounds were analyzed using a Water Xevo G2xs QTof-MS with iclass UPLC system. Our fishing experiments suggested that the immobilized TrkB receptors on CMMN surfaces were able to selectively retain the known binder and could identify compounds from plant extracts. Total seven compounds were identified and all of which have been previous reported to play a role in the therapy of cognitive impartment.

629556 An Adjacent Matrix Method for the Topological Description of a Molecular Structure in Drug Development

Abstract:

With the emergence of new diseases and the progress in medical science, efforts have never stopped to develop efficient medicine for specified diseases and clinical symptoms. Drug properties and clinical performance are essentially decided by its molecular structure. Amongst various methods, data analysis and feature extraction make it feasible to digitize a molecular structure and relate it to its clinical performance. As an effective digitized expression, molecular descriptor has been widely adopted in this field, most of them are obtained by corresponding adjacent matrix, which is used to describe the connection among atoms. However, adjacent matrix is not limited to describe the connection among atoms or functional groups, it can also be used to represent the relationship among self-defined molecular groups flexibly. If an adjacent matrix is established based on atoms, the number of elements could be significantly large, which makes hard to extract the feature structure for particular therapeutic effect and most computation is focused on connections among atoms. If the self-defined molecular groups can be served as basic units for adjacent matrix, computational time will be greatly reduced and more distinct influence of structure on drug properties can be obtained.

In this paper, different adjacent matrix is defined to explore the relationship between drug structure and its performance from multi-scale perspective. According to different partition rules, the structural relationship among atoms, functional groups and self-defined molecular groups can be represented by corresponding adjacent matrices respectively. Then neural network technology is adopted to analyze the connection between drug properties and the topological description, which is transformed from the adjacent matrix. A suitable definition of the adjacent matrix is confirmed to connect drug structure and its performance, on this basis, the contribution of different molecular structure on therapeutic effect, adverse effect and tolerance of the drug can be summarized, which is helpful for property prediction during drug development. Besides, antihemorrhagics drugs are taken as an example for validation purpose. The method proposed in this paper offers an alternative to study the influence of molecular structure on the performance of the drugs and to assist drug developers in designing drug structure.

629479 A Model-Based Approach to Identify Optimal Formulations for the Spray Drying of Pharmaceutical Compounds

Abstract:

Spray drying dispersion (SDD) of active pharmaceutical ingredients (APIs) is proven an effective technique for improving solubility, dissolution and bioavailability of poorly soluble drugs, and it is widely used in pharmaceutical manufacturing. However, preparing suitable formulations for SDD remains a challenging task as it requires identifying proper polymer and solvent blends with desired properties to ensure high solubility and stability of the drug [1]. In current practice, experimental protocols and heuristic approaches are extensively employed for pre-screening small sets of commonly used polymers and solvents for the spray drying of APIs. These methods are time consuming leading to longer development timelines. It is therefore important to develop computational methodologies to reduce the number of time-consuming experiments and enhance the quality and attributes of the drug product.

Over the last decade, a few model-based studies have employed thermodynamic tools to predict solubility and phase stability of given APIs and polymers [2-4]. However, they focus on a limited set of pre-defined mixtures with fixed ingredients and do not fully explore the entire design space of different materials and proportions in order to optimise the final formulation. In this work we present a systematic methodology for identifying optimal API-polymer-solvent formulations with desired physical and chemical properties that are used in the spray drying of drug products. In particular, we employ property-prediction models to estimate solubility, miscibility and the glass transition temperature of a wide range of chemical blends. In addition, we use advanced optimisation models [5] to design improved and environmentally friendly formulations that yield high solubility and stability of the drug. The design methodology is applied to the selection of optimal polymers and solvents that maximise the solubility of naproxen. A ranked list of optimal solutions (blends with different chemicals and compositions) is obtained by introducing integer cut inequalities into the model, and proper phase diagrams of the best mixture are constructed. Finally, the proposed model is incorporated in a polymer and solvent selection workflow that is used to identify better-performing designs and guide experimental work.

[1] M. Davisa, G. Walker, 2018. Journal of Controlled Release 269, 110-127.

[2] Y. Tian, J. Booth, E. Meehan, D. S. Jones, S. Li, G.P. Andrews, 2013. Molecular Pharmaceutics 10, 236–248.

[3] K. Bansal, U. S. Baghel, S. Thakral, 2016. AAPS PharmSciTech 17, 318-327.

[4] K. Lehmkemper, S.O. Kyeremateng, O. Heinzerling, M. Degenhardt, G. Sadowski, 2017. Molecular Pharmaceutics 14, 157-171.

[5] S. Jonuzaj, J. Cui, C.S. Adjiman, 2019. Computers & Chemical Engineering 130, 106518.

624068 Spray-Dried Amorphous Solid Dispersions for Nasal Delivery: Development, Characterization and Performance

Abstract:

INTRODUCTION

Dry powder formulations for nasal drug delivery have advantages over liquid dosage forms, namely the improved stability and increased nasal residence time[1]. For poorly soluble drugs, amorphous solid dispersions (ASDs) are an effective strategy to improve drug solubility, dissolution rate and, consequently, bioavailability[2,3]. They consist of a molecular dispersion of a drug in one or more polymers. ASDs can be prepared by spray-drying and the polymers used can be selected with a dual function: stabilizing the amorphous form and increasing mucoadhesive properties, potentially increasing both dissolution and residence time of the formulation on the nasal cavity.

Given the potential of ASDs for nasal delivery of poorly soluble drugs, the aim of this work was to apply the solubility enhancing properties of ASDs to a poorly soluble model drug (piroxicam) while comparing two particle engineering strategies: spray-drying to obtain particles within the nasal size range of 10-45 μm[1], and agglomeration of spray-dried primary particles into chimeral agglomerates. Both techniques were benchmarked against equivalent physical blends.

METHODS

Piroxicam was selected as model drug, and polyvinylpyrrolidone/vinyl acetate (PVP/VA) and hydroxypropyl methylcellulose E3 (HPMC) as polymers. Spray-drying was performed, using ultra-sonic (USN) and two-fluid nozzle (TFN) to produce microparticles within the nasal size range and primary particles for agglomeration, respectively, in a Büchi model B-290 unit. Chimeral agglomerates were produced by vibrating primary particles in a sieve shaker equipped with 106 μm and 710 μm mesh size sieves. Agglomerates retained on top of the 106 μm sieve were collected. Physical blends were obtained by mixing neat polymer microparticles with piroxicam raw material in a Turbula blender. All formulations were produced at 20% (w/w) drug load.

The powders were characterized concerning their particle size (laser diffraction), morphology (scanning electron microscopy), water content (Karl Fisher) and solid state [differential scanning calorimetry (DSC) and X-ray powder diffraction (XRPD)]. Powders were filled in QUALI-V®-I capsules size 3. Aerodynamic performance was evaluated using a gravimetric 3-stage Andersen Cascade Impactor, equipped with a pre-separator and a 5-L glass expansion, using the active device Miat nasal insufflator. Stage 0, 2, and F were considered, simulating the intranasal passageways (aerodynamic particle size distribution [aerodynamic particle size distribution (aPSD) > 9.0 μm], gastrointestinal tract (aPSD from 4.7 μm to 9.0 μm) and respiratory system (aPSD < 4.7 μm), respectively[4]. The delivered dose during insufflation was determined using a Dosage Unit Sampling Apparatus, and drug quantification by High Performance Liquid Chromatography (HPLC). *In vitro* release testing (IVRT) studies were performed using Franz diffusion cells with a 5mL receptor compartment and a diffusion area of 0.636 cm^2. A dialysis cellulose membrane (MWCO ~ 12,000 Da) was used to separate donor and receptor compartments and simulated nasal fluid (SNF) (8.77 mg/mL NaCl, 2.98 mg/mL KCl, 0.78 mg/mL $CaCl_2.2H_2O$) at a pH of 6.0 was defined as the receptor solution. The receptor compartment was stirred at 600 rpm and maintained at 37°C (assuring 32 °C at the membrane surface), mimicking in vivo conditions[5]. Formulations were applied directly in the donor compartment (5 mg) after weighing it onto a filter paper, followed by the addition of 50 μL of the SNF. Samples were withdrawn for 4 hours and analyzed by HPLC.

RESULTS

PXC microparticles were manufactured by spray-drying at 20% drug load and containing either HPMC or PVP/VA. When using a TFN, a considerable percentage of the particles had sizes below 10 μm. In order to manufacture particles within target particle size range, a 48 kHz USN was used due to its ability to produce larger and more uniform droplets. Spray-dried powders were successfully generated with yields above 39% and Dv50 between 28 and 45 μm. For the chimeral agglomerates, primary particles were manufactured by spray-drying using a TFN. Particle size fell in the desired range for agglomeration[6] (Dv50 between 2 and 3 μm). Using either USN or TFN, particles containing HPMC had higher particle sizes and shriveled morphology, while PVP/VA particles were smaller and had a smooth surface. Chimeral agglomerates were obtained with yields above 80%. These agglomerates have been described as «chimeral» as their size is transient, being reduced during insufflation[6].

DSC and XRPD analysis identified spray-dried microparticles and chimeral agglomerates as amorphous solid dispersions. However, HPMC particles showed recrystallization and melting events on DSC, suggesting higher susceptibility to phase separation and drug crystallization over time[7]. As the blends were manufactured by mixing crystalline drug with amorphous polymer, the resulting product shows a XRPD diffractogram with the amorphous pattern of the polymer and the crystalline peaks characteristic of the drug. Water content measurements identified high water contents for all formulations (>3.5% (w/w)) except for spray-dried microparticles of HPMC (1.55% (w/w)).

Aerodynamic performance studies showed that the fraction potentially retained in the nasal cavity (expansion chamber, pre-separator and Stage 0) was higher than 95% for all formulations (Fig. 1A). Delivered dose was high and reproducible for spray-dried microparticles of HPMC (97.5 ± 1.6 %). For the rest of the formulations, it was lower (51.6 to 75.1 %) and highly variable (standard deviations of 12.3 to 35.5 %), which can be due to the high water content values.

IVRT studies (Fig. 1B) evidenced that both the manufacturing strategy and the polymer have impact on the release behavior of the powder formulations. Spray-dried microparticles and chimeral agglomerates show increased amount of released piroxicam than the corresponding blends, most likely due to the amorphous state of the drug. Additionally, within each formulation manufacturing strategy type, powders containing HPMC showed increased mean percentages of drug released during the duration of the dissolution test (4h). This may be due to the enhanced ability of HPMC to inhibit drug precipitation in a supersaturated solution on the donor compartment, increasing the concentration gradient between both compartments for a longer period of time, and leading to a higher rate of diffusion.

CONCLUSIONS

ASD for nasal delivery were successfully produced using two different polymers at a 20% drug load, by three different manufacturing strategies. Microparticles within the nasal size range showed adequate aerodynamic performance and enhanced drug release properties compared with corresponding blends. For HPMC formulations, chimeral agglomerates, which require an extra manufacturing step, did not present advantages over spray-dried microparticles. Overall, the results support that the HPMC spray-dried microparticles with a 20% drug load would be the lead formulation candidate for further in vitro and in vivo testing, due to the superior and least variable emitted dose and drug release performance. Future work includes mucoadhesion and permeation testing.

References

1. Tiozzo Fasiolo, L, Manniello, MD, Tratta, E, Buttini, F, Rossi, A, Sonvico, F, Bortolotti, F, Russo, P and Colombo, G. Opportunity and challenges of nasal powders: Drug formulation and delivery. *Eur. J. Pharm. Sci.* 2018, 113: 2–17.
2. Vig, B and Morgen, M. Formulation, Process Development, and Scale-Up: Spray-Drying Amorphous Solid Dispersions for Insoluble Drugs. in *Developing Solid Oral Dosage Forms* (Academic Press, 2017). 793–820 doi:10.1016/B978-0-12-802447-8.00030-3.
3. Pozzoli, M, Traini, D, Young, PM, Sukkar, MB and Sonvico, F. Development of a Soluplus budesonide freeze-dried powder for nasal drug delivery. *Drug Dev. Ind. Pharm.* 2017, 43: 1510–

1518.
4. Copley, M and Kippax, P. From actuation to deposition: Particle sizing techniques for characterizing nasal drug delivery systems. *Inhalation* 2012, 4: 12–16.
5. Miranda, M, Pais, AACC, Cardoso, C and Vitorino, C. aQbD as a platform for IVRT method development – A regulatory oriented approach. *Int. J. Pharm.* 2019, 572: 118695.
6. Colombo, P, Santi, P, Bettini, R, Catellani, PL, Artusi, M and Sacchetti, C. Powder for nasal administration of drugs. 2005, .
7. Duarte, Í, Santos, JL, Pinto, JF and Temtem, M. Screening Methodologies for the Development of Spray-Dried Amorphous Solid Dispersions. *Pharm. Res.* 2015, 32: 222–237.

Figure 1 – (A) Aerodynamic profile of powder formulations. **(B)** In vitro drug release profile of powder formulations. SDM – spray-dried microparticles; CA – chimeral agglomerates; PVP/VA - polyvinylpyrrolidone/vinyl acetate; HPMC - hydroxypropyl methylcellulose.

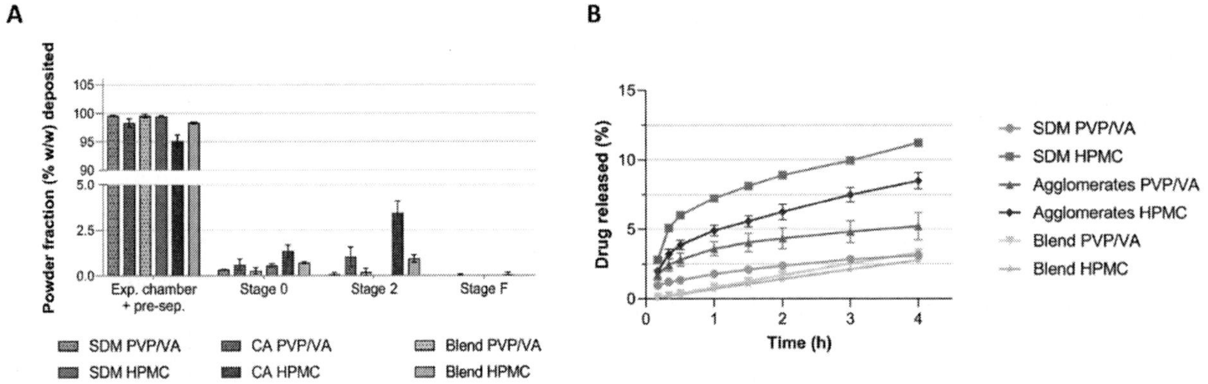

627703 Paliperidone: In Vitro Controlled Release from Long-Acting Prodrug Depot Suspension

Abstract:

One of the main goals of formulation scientists in the last several years is to ensure the delivery of BCS II and IV drugs (90% of all pipeline drugs [1]) at a reasonable rate and extent of bioavailability [2]. A possibility to overcome the negative clinical effects caused by poor water solubility is to increase the surface area, solubility, and wettability of powder drug particles by particle size reduction [3]. This trend of shifting pharmaceutical formulations to an ever-smaller scale has led to micro- and nano-formulations, such as intramuscular depot suspensions. These have a number of advantages over conventional oral dosage forms, including increased bioavailability, sustained/controlled drug release over months and the ability to maintain drug concentrations within the therapeutic range for a long time period. In particular, these benefits contribute to increasing patients' adherence to the pharmacotherapy, which is currently a frequently mentioned issue. It has been estimated that the cost to the US health care system in 2012 caused by patient nonadherence was about $200 billion [4].

The present work is focused on the intramuscular prodrug (paliperidone palmitate) depot nano-suspension, which is used in common practice as an intramuscular long-acting medication for schizophrenia treatment. Although this form has many advantages over the previous oral dosage form (sustained release, improvement of patients' adherence and quality of life), its disadvantage is the lack of IVIVC correlations and the absence of a detailed description of the parent drug (paliperidone) release mechanism from the drug formulation. The lack of IVIVC studies compared to conventional oral dosage forms is a general problem with depot formulations development, and a major barrier to their widespread use [5]. This, together with a lack of understanding of the *in vivo* release mechanism, often leads, in a generic drug development of such formulations, to the success in dissolution testing followed by failure in bioequivalence evaluation, which causes considerable time and financial losses.

The result of this work is an experimentally validated mathematical model describing the *in vitro* release of paliperidone from a nano-suspension of paliperidone palmitate (prodrug). The mechanism of paliperidone *in vitro* release is a reactive dissolution, i.e., dissolution of solid particles of paliperidone palmitate to the surrounding medium followed by conversion to paliperidone in the liquid phase. This mechanism mimics the *in vivo* release process that takes place after the intramuscular administration, where at the injection site a depot of prodrug particles is formed, from which the prodrug gradually dissolves into the tissue fluid and is converted to the parent drug by naturally occurring enzymes. In contrast to conventional dissolution tests, the proposed reactive dissolution model thus reflects not only the solid paliperidone palmitate particles dissolution kinetics but also the kinetics of its subsequent conversion. Such model can thus serve as a springboard for subsequent IVIVC correlation development. In its current form, the model can solve two types of tasks. The first, called as the "forward problem", is the prediction of the release profile of paliperidone based on the initial particle size distribution (PSD) of the suspension. The second, the so-called "inverse problem", can predict the initial PSD that will lead to obtaining the desired release profile. Specifically, the ability to solve the inverse problem could be used in the suspension dosage forms formulation strategy, where particle size modification steps are usually included in manufacturing, but the relationship between PSD and drug release rate is typically determined experimentally and often covers only a limited subset of the parametric design space.

The mathematical model describes the physical dissolution from different particle size classes of the suspension by population balance equation (Randolph & Larson; 1971) coupled with Nernst-Brunner equation, which depicts the dissolution kinetics of isolated spherical crystal. The mass transfer coefficient in a stirred vessel containing the depot suspension was obtained from Sherwood number calculated on the basis of the correlation proposed by Armenante & Kirwan (1989) suitable for particles of the order of few micrometers. Reynolds number was determined from particle diameter, kinematic viscosity, and energy dissipation rate. The dissolution kinetics was supplemented with the kinetics of paliperidone palmitate hydrolysis to paliperidone, which was measured experimentally. In order to reduce the time required to perform the experimental part, the hydrolysis of paliperidone palmitate was performed with hydroxide and not with enzyme. The mechanism of the basic hydrolysis (S_N2 nucleophilic substitution) is known from the literature and the rate constant of prodrug conversion was determined from the experimental data. Due to the basic environment (pH ~11), mild degradation of paliperidone was observed in the experiments, which is also included in the developed model. The solution of the inverse problem is based on the same equations, but the model works iteratively using the Nelder-Mead optimization algorithm in order to find such combination of individual size fractions in a multimodal PSD that minimizes the difference between the desired and actual release profiles.

To validate the model in terms of solving the forward problem, a custom dissolution apparatus was designed, consisting of a 250 mL glass vessel with controlled agitation and temperature maintenance. A dissolution medium was added to this vessel into which the paliperidone palmitate suspension was being dissolved. The dissolution medium contained a reagent (lithium hydroxide) which ensured the conversion of paliperidone palmitate to paliperidone in the liquid phase. Although the conversion of paliperidone palmitate *in vivo* takes place enzymatically, basic hydrolysis has been used, which can be carried out at higher temperatures (increasing and overall process rate), has a known mechanism and is less sensitive to environmental conditions. To measure the time evolution of the concentration of both substances, an HPLC (Agilent 1100) method was developed. The reliability and accuracy of the experimentally measured data were verified on the basis of criteria set by the ICH guidelines (CDER, US FDA).

Validation of the ability to solve the inverse problem has so far been performed on a different system than paliperidone, mainly in terms of time savings as paliperidone palmitate dissolution rate is very low. The experiments were performed with a suspension of potassium chloride (KCl) crystals. Because KCl dissolves very rapidly in aqueous media, the real-time dissolution kinetics is difficult to be measured. Therefore, the experiments were performed in a mixture of water and isopropanol (3:10 by volume), which slowed down the dissolution rate. The real-time measurements of PSD evolution and dissolution kinetics were performed in the flow cell of a static light scattering instrument (Horiba Partica LA-950). The flow cell served as a dissolution tank in which a conductometer (Mettler Toledo SG23) was inserted to measure the KCl concentration.

REFERENCES

1. Benet, L., In *EDAN*, Leuven, Belgium, March 2017.

2. Lipp, R., Major advances in oral drug delivery over the past 15 years. American Pharmaceutical Review 2013, 16.

3. Subramaniam, B.; Rajewski, R. A.; Snavely, K., Pharmaceutical Processing with Supercritical Carbon Dioxide. Journal of Pharmaceutical Sciences 1997, 86 (8), 885-890

4. IMS Institute for Healthcare Informatics Avoidable costs in US health care. 2013

5. Chavda, V., et al. (2016). "In Vitro–In Vivo Correlation" Research & Reviews: A Journal of Drug Formulation, Development and Production

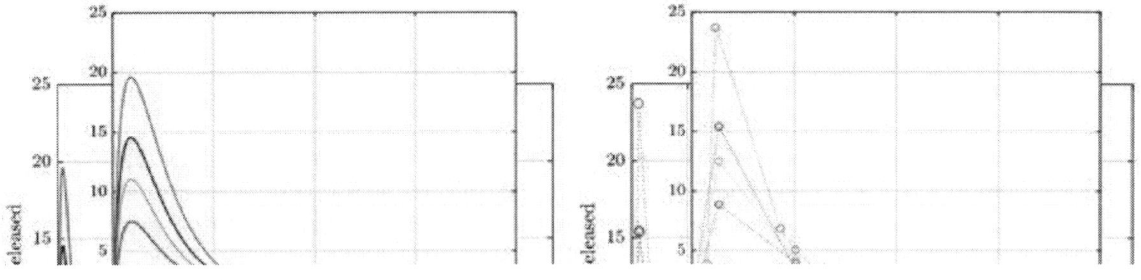

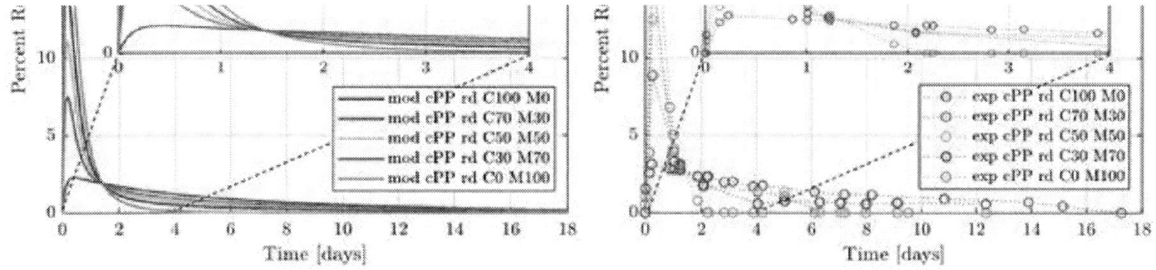

Fig. 1: Model (left) and experimental (right) results for paliperidone palmitate release profiles obtained for two suspensions of different initial particle size distribution, referred to as: CRUDE (d_{50} = 12um) and MILLED (d_{50} = 1um), and their combinations in ratios described in the legend

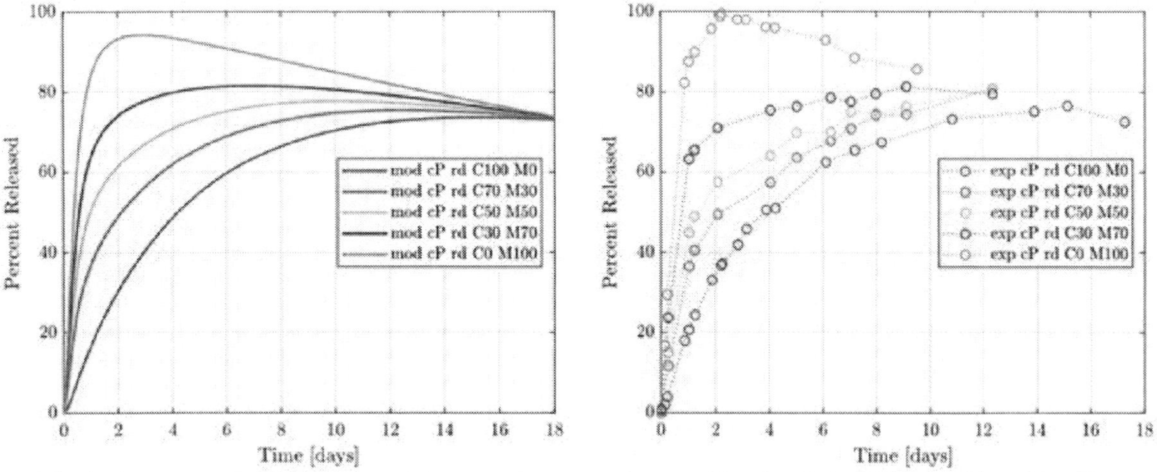

Fig. 2: Release profiles the same as in Figure 1, but corresponding to paliperidone instead of paliperidone palmitate

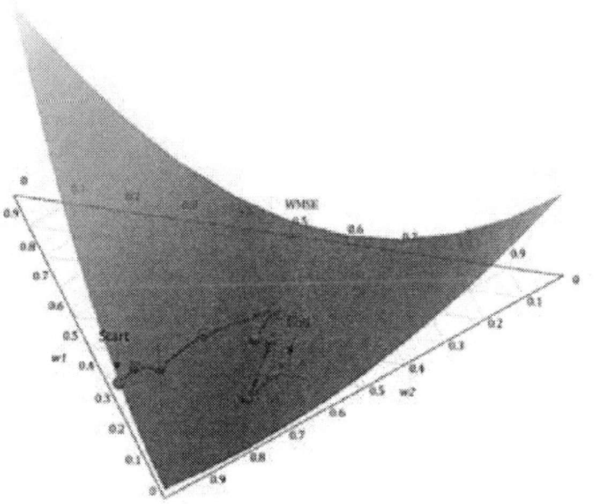

Fig. 3: Optimization by Nelder-Mied algorithm in search of an ideal PSD composed of three monomodal PSDs, which would ensure the required release profile

626725 Use Marker Pen to Constantly Release Tetrodotoxin at Nanogram Level

Abstract:

Tetrodotoxin (TTX) is an effective non-protein specific blocker of voltage-gated sodium (NaV) channels. TTX is 1000-fold more potent than clinically used anesthetics such as bupivacaine and lidocaine. In addition, TTX does not cause myotoxicity or neurotoxicity, has no affinity with the myocardial NaV channels, and does not penetrate the blood-brain barrier to cause drug addiction. All these features make TTX an attractive candidate for painkiller design. However, the principal reason why TTX has not yet achieved clinical application is its systemic toxicity. The TTX overdose causes neural blockade and muscular weakness resulting in diaphragmatic paralysis leading to respiratory arrest and death. Therefore, the final and most critical milestone of TTX clinical practice is to precisely control the dose of TTX to take advantage of its therapeutic effects but avoid its systemic toxicity. Inspired by the marker pen, which stores the water-based ink in the spongy core and always releases a constant amount of ink from the tip for writing, we explored the marker pen for TTX controlled release. We have verified that the marker pen can encapsulate water-soluble drugs in the pen core by absorbing the drug aqueous solution, and release a constant amount of drug per unit length of the line drawn with the pen tip. In addition, the drug concentration in the filling drug aqueous solution can be infinitely diluted to achieve the constant release of drug at the nanogram level or less (such a low level makes TTX super safe to use). The marker pen is equipped with a pen stopper to prevent leakage of drug. We also demonstrated that the drug-filled pen has a super-long shelf life as evidenced that the amount of drug released from the pen 6 months after filling is the same as that of a newly filled pen. One practical application of the marker pen filled with TTX is a topical anesthetic, which is used to treat skin pain.

AUTHOR INDEX

AIChE
120 Wall Street, FL 23
New York, NY 10005-4020

CURRAN ASSOCIATES INC.
proceedings
.com

ISBN 978-1-7138-5719-8
9 781713 857198